W9-CPQ-978

SAMPLING AND CALIBRATION FOR ATMOSPHERIC MEASUREMENTS

A symposium sponsored by
ASTM Committee D-22 on
Sampling and Analysis
of Atmospheres
Boulder, CO, 12–16 Aug. 1985

ASTM SPECIAL TECHNICAL PUBLICATION 957
John K. Taylor, National Bureau
of Standards, editor

ASTM Publication Code Number (PCN)
04-957000-17

 1916 Race Street, Philadelphia, PA 19103

Library of Congress Cataloging-in-Publication Data

Sampling and calibration for atmospheric measurements.

(ASTM special technical publication; 957)
"ASTM publication code number (PCN) 04-957000-17."
Papers presented at the Conference on Sampling and Calibration for Atomspheric Measurements.
Includes bibliographies and index.
1. Air quality—Measurement—Congresses.
2. Air—Pollution, Indoor—Measurement—Congresses.
I. Taylor, John K. (John Keenan), 1912–   . II. ASTM Committee D-22 on Sampling and Analysis of Atmospheres.   III. Symposium on Sampling and Calibration for Atmospheric Measurements (1985: Boulder, Colo.)
IV. Series.
TD890.S26   1987   628.5'3   87-12439
ISBN 0-8031-0955-5

Copyright © by AMERICAN SOCIETY FOR TESTING AND MATERIALS 1987
Library of Congress Catalog Card Number: 87-12439

NOTE
The Society is not responsible, as a body,
for the statements and opinions
advanced in this publication.

Printed in Baltimore, MD
August 1987

Foreword

The symposium on Sampling and Calibration for Atmospheric Measurements was held in Boulder, Colorado, 12–16 August 1985. The symposium was sponsored by ASTM Committee D-22 on Sampling and Analysis of Atmospheres. John K. Taylor, National Bureau of Standards, Richard G. Melcher, Dow Chemical Company, and Harry L. Rook, National Bureau of Standards, presided as symposium chairmen. John K. Taylor is editor of this publication.

Related
ASTM Publications

Quality Assurance for Environmental Measurements, STP 867 (1985), 04-867000-16

Toxic Materials in the Atmosphere: Sampling and Analysis, STP 786 (1982) 04-786000-17

Sampling and Analysis of Toxic Organics in the Atmosphere, STP 721 (1981), 04-721000-19

Air Quality Meteorology and Atmospheric Ozone, STP 653 (1978), 04-653000-17

Calibration in Air Monitoring, STP 598 (1976), 04-598000-17

A Note of Appreciation
to Reviewers

The quality of the papers that appear in this publication reflects not only the obvious efforts of the authors but also the unheralded, though essential, work of the reviewers. On behalf of ASTM we acknowledge with appreciation their dedication to high professional standards and their sacrifice of time and effort.

ASTM Committee on Publications

ASTM Editorial Staff

Helen M. Hoersch
Janet R. Schroeder
Kathleen A. Greene
Bill Benzing

Contents

Introduction

This publication contains the texts of a series of papers presented at the Conference on Sampling and Calibration for Atmospheric Measurements sponsored by ASTM Committee D-22 on Sampling and Analysis of Atmospheres. This conference was held at the University of Colorado, Boulder, Colorado, during 12–16 August 1985. It was the seventh in a series of biannual conferences to advance the state of the art of atmospheric measurements. Previous conferences were devoted to the following subjects.

Conference Date	Title (Publication Date)
1973	Instrumentation for Monitoring Air Quality ASTM STP 555 (1974)
1975	Calibration in Air Monitoring ASTM STP 598 (1976)
1977	Air Quality Meteorology and Atmospheric Ozone ASTM STP 653 (1978)
1979	Sampling and Analysis of Toxic Organics in the Atmosphere ASTM STP 721 (1981)
1981	Toxic Materials in the Atmosphere, Sampling and Analysis ASTM STP 786 (1982)
1983	Quality Assurance for Environmental Measurement ASTM STP 867 (1985)

The 1983 Conference on Quality Assurance for Environmental Measurements focused on the general topic of data quality. Throughout that conference, the question of the reliability calibration and the relevance of samples frequently was called to attention. Accordingly, it was decided to devote the 1985 conference to a comprehensive discussion of these topics. The papers discuss them from four points of view: General Aspects, Indoor Air, Ambient Air, and Workplace Atmospheres.

The Committee D-22 sponsored conferences have provided a forum for the presentation of state-of-the-art papers in selected areas and the opportunity for both formal and informal discussions by the attendees. The publication of the papers serves to extend the benefits of the conference to a wider audience.

John K. Taylor
Coordinator for Chemical Measurement Assurance and Voluntary Standardization, Center for Analytical Chemistry, National Bureau of Standards, Gaithersburg, MD; editor.

General Topics

Byron Kratochvil[1]

General Principles of Sampling

REFERENCE: Kratochvil, B., **"General Principles of Sampling,"** *Sampling and Calibration for Atmospheric Measurements, ASTM STP 957,* J. K. Taylor, Ed., American Society for Testing and Materials, Philadelphia, 1987, pp. 5–13.

ABSTRACT: In atmospheric measurements, the sampling operation is often a major source of error. This error may be reduced by consideration of the distribution pattern of the substance being measured, and the statistical principles of that pattern. From this information, guidelines as to the number, size, and location of the samples can be formulated. These guidelines should be tempered, however, with judgment and experience.

KEY WORDS: sampling plans, sampling designs, sampling errors, statistical sampling, cost benefit analysis in sampling, air quality, calibration, sampling, atmospheric measurements

The collection of an atmospheric sample for the quantitative or qualitative evaluation of a physical or chemical property is often the most difficult part of the overall process. Ensuring that the increment collected is a valid representation of what the designer of the data intended to be collected is even more difficult. It is therefore worthwhile to consider the sources of the uncertainties introduced into atmospheric measurements by the sampling operations. This paper covers some of the basic principles of sampling from a statistical and practical point of view. Specific applications or sampling systems for atmospheric measurements will not be discussed. Emphasis will be placed instead on the basic design of overall sampling strategy.

Setting Up a Sampling Design

To put sampling into context with the overall analysis, it is useful to consider the sequence of steps involved in the collection of data. This sequence, outlined in Fig. 1, consists of five major actions: model, plan, sample, measure, and evaluate. The model determines the questions to be answered by the study. It defines the goals of the work. The plan sets out the means whereby these goals are achieved. The planning stage includes selection of the analytical methodology to be used, the resources and manpower to be assigned, and the time frame. The plan should also specify the number and size of the samples to be collected and the sampling locations. Where previous information on the population is unknown, the plan may also require a preliminary sampling and measurement to assess the variability of the population. This is followed by the sampling operations, in which field samples are collected and, if necessary, refined to test portions. Next comes the measurement, which involves preparation of the test portion for analysis by dissolution, extraction, and so on, followed by the data acquisition. Lastly, the best value is selected and its precision and

[1] Professor of Chemistry, Department of Chemistry, University of Alberta, Edmonton, Alta., Canada T6G 2G2.

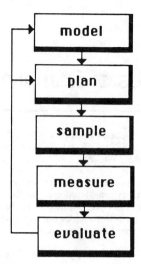

FIG. 1—*Sequence of operations in an overall analytical process.*

accuracy are estimated (evaluation). On the basis of the results, the validity of the model is assessed, revised, if necessary, and the operations repeated.

All these steps affect, and are affected by, the sampling component. In the establishment of the model, such questions must be answered as: "What is the extent of the population to be sampled?" "How much speciation is required?" "Is only an average or integrated composition in time or space needed, or must the variability be assessed?" "What level of precision is needed?" "What are the limitations in time, manpower, and money that constrain other decisions?"

In the planning step, it is important to provide a comprehensive sampling plan that includes information on sample increment size, number, and location, the extent to which individual increments may be combined (composited), and a written protocol that is clearly understood by those responsible for sample collection. It is valuable to involve in the planning step as appropriate the analyst, a statistician, and the end user of the data. The written protocol might include such information as when, where, and how to collect the sample increments; on-site criteria for rejection of foreign matter; descriptions of the apparatus and sample containers to be used, with instructions as to their cleaning and handling so that contamination before or after sample collection is avoided; sample labeling, recordkeeping, and auxiliary material to be included; and sample storage conditions or use of preservatives. It is important that the individual doing the sampling operation be properly trained in the use of the apparatus and in potential sources of error or bias, and that he or she appreciate the significance of the steps necessary to avoid compromise of the sample. In summary, the written protocol should answer the news reporters' old questions of what, where, when, how, and by whom.

Several types of sampling plans may be considered. Questions that may be asked at this stage include: "Should a random or systematic sampling pattern be followed?" "Should sampling be continuous, intermittent, or spot?" and "Should a stratified sampling design be used?" These questions will be considered later. The use of automated sample collectors may also need to be evaluated as to type, efficiency, cost, and availability.

Errors from Sampling

Sampling errors may be classified as either random or systematic. Random errors are errors that vary in direction and magnitude; they can be treated by statistical methods. Systematic errors cause a measured value to fall always above or below the true value. They introduce a bias into the results. Systematic errors cannot be treated by statistics.

Random error can never be eliminated, but its magnitude may be reduced by careful work. It may originate in the variability of the material being sampled or in the sampling operation itself. So long as the population being sampled is not completely homogeneous, some uncertainty will be introduced into an estimate based on analysis of a limited portion of the population, even if the analysis of the sample were error-free.

Systematic errors, or bias, may arise from many sources, and constant vigilance is necessary to keep them to a minimum. Sampling bias may be introduced by faulty sampling design, by improper sampling operations, or by sample changes before measurement. An automatic sampler for particulates in air, for example, may discriminate in an unexpected way according to particle size, mass, or other property. Or, additional measurements of time, temperature, flow rate, pressure, etc. needed to calculate concentrations may have been made incorrectly. Samples also may undergo deterioration during collection, transport, storage, or preparation for analysis. Components in the sample may interact with other constituents of the sample, or with transfer lines or containers during collection and storage. Reaction of sample components with oxygen, water, or carbon dioxide in the air is common. Loss of sample by volatility of water or other substances is possible. Also, samples may undergo stratification or segregation on storage owing to differences in density, particle size, and so on. All these factors must be watched for and kept under control if bias is to be held to a minimum.

Errors related to the model include improper handling of outliers, such as inclusion of foreign objects in a sample or mistaken rejection of substances that are not foreign; failure to recognize hot spots or unusual distributions in the population; use of improper increment sizes so that particle size causes bias; or absence of an adequate chain of custody for the samples.

Outliers are often a problem when the distribution pattern of the sought-for substance is unknown because a suspect value may be a hot spot or nugget and, therefore, a valid part of the population. Distinguishing between a result that should be rejected owing to gross error and a valid data point can be assisted somewhat by considering the following conditions. First, if the general distribution of the sought-for material is known to be at least approximately Gaussian, statistical tests such as the Q test [1] or a related approach [2] can be applied. In the Q test, a Q_{obs} is calculated by the relationship

$$Q_{obs} = \left| \frac{\text{suspect value-nearest value}}{\text{range}} \right|$$

where "nearest value" is that value numerically closest to the suspect value. If Q_{obs} exceeds a tabulated value, Q_{tab}, for a given small number of measurements ($n = 3$ to 10), the suspect value may be rejected at the specified confidence level. Examples of table values at the 90% confidence level [1] are: (n, Q_{tab}) 3, 0.94; 4, 0.76; 5, 0.64; 6, 0.56; 7, 0.51; and 8, 0.47. In a more general procedure, Natrella applies the same approach but for $n = 3$ to 7, and provides a table of Q_{tab} values for upper percentiles ranging from 0.70 to 0.995 [2]. Second, if an error or blunder in the sampling operation is identified, the value may be rejected. And third, if neither of the preceding points applies, yet the result is on the basis of experienced judgment clearly unreasonable, rejection is appropriate.

Basic Statistical Considerations for Sampling

For random errors, the best estimate of the uncertainty of the average is the standard deviation s, or the variance s^2. The overall variance, s_o^2, for a series of analytical operations is the sum of the variances for the individual steps. For an analysis that requires collection of a field sample, subsampling of that field sample, and analysis of the subsample, the overall variance is given by

$$s_o^2 = s_s^2 + s_{ss}^2 + s_a^2$$

To increase the precision of a measurement requires a reduction in s_o. To accomplish this, all of the contributing variances must be considered. It does no good to reduce all but one of the contributing variances if that one is significant. For example, if the standard deviation of sampling, subsampling, and analysis, (s_s, s_{ss}, and s_a, respectively) are all 2%, s_o is 3.5%. Reduction of both s_{ss} and s_a by a factor of 10 from 2 to 0.2% yields an s_o of 2.02. Further reduction of s_{ss} or s_a will not improve s_o to a measurable extent. Youden proposed that once a component contributing to s_o had been reduced to less than a third of s_o, it is neither time nor cost effective to try to reduce further the standard deviation of that component [3]. Replication of any operation will of course reduce the standard deviation of a measurement based on that operation. With knowledge of the standard deviation for a single sample, measurement, or other contributing factor to the overall standard deviation, the improvement provided by increasing the number of replicates is given by

$$s_{o,m}^2 = \frac{s_s^2}{x} + \frac{s_{ss}^2}{xy} + \frac{s_a^2}{xyz} \tag{1}$$

where $s_{o,m}$ is the overall standard deviation of the mean, x is the number of sample increments, y is the number of subsamples taken from each sample increment, and z is the number of analyses performed on each subsample [4].

Some Statistical Distributions Found in Nature

Practically all the common statistical tests used in the evaluation of analytical measurements apply only to Gaussian distributions. Gaussian, or normal, distributions are characterized by symmetry about the mean, or average, and by small deviations from the mean being more frequent than large ones. This leads to the distribution having a familiar bell-shaped curve. The width of the curve is determined by the value for the standard deviation of the data set. Other distributions are frequently encountered in nature, and it is important to determine whether the data set is Gaussian prior to applying conventional statistical tests. This can be done by use of the Kolmogorov or other tests [5,6]. The Kolmogorov test, which can be performed either graphically or numerically [7] and which is applicable to small data sets, involves tabulation of a cumulative frequency distribution for a set of results. The largest difference between the observed distribution and that expected for a Gaussian distribution is then computed. This maximum difference is compared with tabled critical values; if the observed difference is larger than the table value, the null hypothesis can be rejected.

If the distribution is other than Gaussian, it may be possible to transform the data to fit a Gaussian distribution and then apply the usual tests. Other distributions that may be encountered in Nature include the log normal, binomial, Poisson, and negative binomial [6,8]. Examples include: for log normal, the distribution of a metal in a low-grade ore deposit or the distribution of molecular chain lengths in a crude oil; for binomial, the distribution of defective items in a production line or the distribution of a substance in a mixture of two kinds of particles; for Poisson, the distribution of particulates in the atmosphere or the distribution of wear metals in a lubricating oil; and for negative binomial, the clumped distribution of point source mold growth in a bin of grain or of weed or insect infestation in a field.

Systematic versus Random Sampling

Statisticians carefully point out that only if all portions of a population have an equal chance of being selected during sampling can valid statistical conclusions be drawn about the population. This requires use of some type of random sampling plan. Typically, the area or volume to be sampled is divided on paper into numbered units and the units to be sampled are selected from a table of random numbers. Such tables are available from a variety of books or from computer programs [2]. In this way, the possibility of bias, either conscious or unconscious, is reduced. Random sampling is often difficult and costly, however. Field workers may not appreciate the importance of collecting a sample at the specified location or time if it is not readily accessible or convenient. Also, it is not easy to use automated equipment to collect samples in a random pattern.

The alternative to random sampling is systematic sampling. In systematic sampling, sample increments are collected at regular intervals in space or time. Systematic sampling has the advantages that sample increments are simpler to collect, either manually or with automatic apparatus, are less subject to errors in collection, and are distributed more evenly over the population. The major disadvantage is that if the population contains a periodic variation in the substance being measured, and the sampling frequency or period coincides with the population variation or some multiple of it, the sample may be biased. Error from this source can be reduced by changing the sampling frequency or spacing at least once during the sampling operation and testing for differences between the sampling frequencies. When using systematic sampling, it is critical to reduce bias by selection of the time or location of the initial sample in the series by a random process. This maintains an element of randomness in the set.

Sampling Stratified or Segregated Materials

If the population under study is known or suspected of being segregated in some way, stratified sampling should be considered. In this procedure, the population is divided into parts, or strata, each as uniform in the component of interest as possible. The strata do not need to be equal in size. Each stratum is then sampled independently, and the average and standard deviation determined. From knowledge of the size of each stratum and its composition, the overall composition and variability can be calculated.

The advantage of stratified sampling is that differences between the averages for the strata do not contribute to the sampling uncertainty; the sampling standard deviation arises only from within-stratum variations. This means that stratified sampling never gives poorer precision than simple random sampling, but may give significantly better precision, depending on the uniformity of stratum composition for the sought-for component. For maximum efficiency, the number of initial sample increments should be proportional to the size of each stratum. In subsequent sampling, those strata showing greater variability should be sampled more intensively.

In deciding how many strata to establish, and how large each should be, draw on past information on the population. If past information is lacking, it is necessary to use judgment and past experience. Time and cost factors may also be major considerations.

Stratification may take place either in space or in time. When sampling air particulates near a highway or a factory, for example, it may be useful to stratify sample collection into time periods determined by hours of heavy morning and evening traffic or plant operation.

Estimation of the Number of Sample Increments

When sampling atmospheres, the sample is often collected on an adsorbent or filter over a relatively long period of time, such as a working day or 24 h. Analysis of the collected material provides an integrated value over the given time period. A second method is to use an instrument that gives a continuous reading of the concentration of the monitored substance. For the first method, it is of interest to know how many collection devices per given area are needed to provide a desired precision. In the second case, it may be worth knowing whether the instrument can be operated for a few short periods a day at a single location, or whether multiple locations must be sampled. Once the variability of the population (and of the sampling and measurement operations) has been established, it is possible to calculate how many samples must be taken to achieve any desired level of precision. For example, the time periods in a working day, or the work areas in a plant, may be considered as discrete units. Then the variance of the mean s_m^2 is given by

$$s_m^2 = \frac{s_b^2}{N_b} = \left(\frac{N - N_b}{N}\right) + \frac{s_w^2}{N_b N_w} + \frac{s_z^2}{N_b N_w N_z}$$

where N is the total number of units, N_b is the number of units sampled, N_w is the number of sample increments taken per unit, and N_z is the number of analyses performed per sample increment according to ASTM Recommended Practice for Choice of Sample Size to Estimate the Average Quality of a Lot or Process (E 122-72) and Refs 4 and 9.

If separation of the population into arbitrary units is not feasible, then the number of sample increments, N, to be collected for a given level of sampling uncertainty, R, can be calculated if the average composition, $\bar{x}$, and degree of heterogeneity of the population for the sought-for substance, s_s, is known [9]. The relationship is

$$N = \frac{t^2 s_s^2}{R^2 \bar{x}^2} \times 10^4 \qquad (2)$$

Here, s_s and $\bar{x}$ are the known sampling standard deviation and average values for the population, R is the relative sampling standard deviation (in percent) that is acceptable, and t is the student t value that applies to the desired level of confidence. The relative standard deviation, R, is given by $100 \, s_s/\bar{x}$. The value for t is obtained from a t-table, available in books on chemical analysis or statistics [10]. Initially, t can be set at the value for 95% confidence limits, 1.95, and an initial estimate of N calculated. The t value for this N can then be substituted and the calculation repeated until N is constant.

The preceding expression applies only to a Gaussian distribution. For a Poisson distribution, s_s^2 is of the same order of $\bar{x}$, and Eq 2 simplifies to

$$N = \frac{t^2}{R^2 \bar{x}} \times 10^4$$

If the distribution is negative binomial, inclusion of an index of clumping k is necessary, and Eq 2 becomes

$$N = \frac{t^2}{R^2}\left(\frac{1}{\bar{x}} + \frac{1}{k}\right)(10^4)$$

Values for k and $\bar{x}$ must be estimated from previous measurements.

Cost Benefit Analysis in Sampling

When devising a practical sampling plan, compromises are often necessary. Collection of the ideal number and variety of samples may not be feasible owing to limitations of time, money, or manpower availability. Questions that are often raised are: "What is the best sampling plan for a given cost?" or "How much will it cost to obtain data of a given precision?" Answers to these questions can be found if information on the precision and cost of the various sampling and analytical operations is known or can be estimated [11].

The expressions to calculate the first case, that of the optimum plan for a given total cost, c, are

$$x = c\sqrt{s_s^2/c_s}/(\sqrt{s_s^2 c_s} + \sqrt{s_{ss}^2 c_{ss}} + \sqrt{s_a^2 c_a})$$

$$y = \sqrt{(s_{ss}^2 c_s)/(s_s^2 c_{ss})}$$

$$z = \sqrt{(s_a^2 c_{ss})/(s_{ss}^2 c_a)}$$

where x is the number of sample increments to be collected, y is the number of subsamples prepared from each increment, and z is the number of analyses performed on each subsample. Use of this equation requires careful attention to rounding errors, however.

For example, suppose the standard deviation, s, and costs, c, for the sampling, subsampling, and measurement steps of a method were: $s_s = 0.0026$, $s_{ss} = 0.0028$, $s_a = 0.0013$, $c_s = \$30$, $c_{ss} = \$120$, and $c_a = \$80$.[2] If the total amount to be spent were \$400, then, rounding to the nearest integer, $x = 3$, $y = 1$, and $z = 1$. This yields $c = (3 \times 30) + (3 \times 120) + (3 \times 80) = \$690$ and an overall standard deviation, s_o, of 0.002. (The sum is greater than \$400 owing to rounding.) Examination of the data shows that the values of both x and y must be reduced to unity if the cost is to be held below \$400. The best one can do to stay under \$400 is $x = 1$, $y = 1$, and $z = 3$. Now $c = \$390$ and $s_o = 0.004$.

For the case where the minimum cost to attain a predetermined overall standard deviation is to be determined, the equations are

$$x = \sqrt{s_s^2/c_s}\ (\sqrt{s_s^2 c_s} + \sqrt{s_{ss}^2 c_{ss}} + \sqrt{s_a^2 c_a})/s_o^2$$

$$y = \sqrt{(s_{ss}^2 c_s)/(s_s^2 c_{ss})} \qquad\qquad (3)$$

$$z = \sqrt{(s_a^2 c_{ss})/(s_{ss}^2 c_a)}$$

Using the same values for the individual standard deviations and costs as in the previous

---

[2] Data from Dean Wallace, Alberta Research Council, Edmonton, Alta., Canada, 1985.

example, and supposing an overall standard deviation not exceeding 0.002 was wanted, solution of the equations gives $x = 7$, $y = 1$, and $z = 1$ for a total cost of $1610. Insertion of these values into Eq 3 gives $s_o = 0.0015$.

Again, errors caused by rounding must be watched for. In this example, a decrease of x from 7 to 4 will increase s_o to 0.003, just at the acceptable value, but will reduce the overall cost to $920.

Sampling for Identification

Sometimes samples must be collected for qualitative rather than quantitative purposes. For example, the vapors released by a chemical spill from a highway or railroad accident, from a plant, or from a disposal site may need to be identified for control or compliance purposes. In another instance, the head space gas in a waste or other container may need to be sampled. Such problems require special attention to ensure that the analyzed sample adequately represents the population of interest.

An increasingly useful method for qualitative scanning of atmospheres is the application of "sniffer" instruments such as gas chromatographs, infrared spectrophotometers, or mass spectrometers with inlets that operate at atmospheric pressures. These instruments allow essentially instantaneous on-site sampling, analysis, and, often, identification of trace components in the atmosphere. Experience with a specific technique is necessary for proper interpretation of the data, and verification with standards or known substances as well as background testing are important components of careful work. Provision for required ancillary services and for ensuring that the instruments are sufficiently rugged for field work often results in these instruments being costly to purchase and operate.

In head-space analysis, where a sample is pumped through an analyzer or a collector, attention must be given to the rate at which the gaseous components reach equilibrium with the remainder of the system under study. Too rapid a pumping rate may bias the results toward those components that equilibrate more rapidly. Errors of this kind can be tested for by varying the pump rate and by watching for changes in apparent composition with time.

Conclusions and Comments

Anyone setting up a program for sampling of the ambient atmosphere should read the planning guide provided in ASTM Recommended Practice for Planning the Sampling of the Ambient Atmosphere (D 1357–82) for general principles and for guidelines on the selection of sites. Also useful is ASTM Recommended Practices for Sampling Atmospheres for Analysis of Gases and Vapors (D 1605–60(1979)) that includes methods with and without concentration of the sought-for components. A related standard, ASTM Practice for Sampling Atmospheres to Collect Organic Compound Vapors (Activated Charcoal Adsorption Method) (D 3686–84), outlines a procedure for collection of organic compound vapors by adsorption on activated charcoal.

Because of the heterogeneity of the atmosphere, particularly near point sources of possible contaminants, it is necessary to use a sufficient number of sampling points, to locate them at sites selected to minimize bias, and to sample frequently enough, or long enough, to average out short-term fluctuations. Reduction of sampling error to an acceptable level almost always requires sufficient preliminary measurements to establish the variability of the components sought in space and time. With this data in hand, an experienced analyst can devise a valid sampling plan with reasonable cost and effort.

Acknowledgments

The author wishes to acknowledge many useful discussions with John Taylor of the National Bureau of Standards Center for Analytical Chemistry and with Dean Wallace of the Alberta Research Council. Both contributed valuable suggestions that have been incorporated into this article. The assistance of Annabelle Wiseman is also gratefully acknowledged.

References

[1] Dean, R. B. and Dixon, W. J., *Analytical Chemistry*, Vol. 23, 1951, pp. 636–638.

[2] Natrella, M. G., *Experimental Statistics*, National Bureau of Standards Handbook 91, U. S. Government Printing Office, Washington, DC, Aug. 1963, pp. 17–2 and 17–3.

[3] Youden, W. J., *Journal Association of Official Analytical Chemists*, Vol. 50, 1967, pp. 1007–1013.

[4] Hackler, W. C., Clatfelter, T. E., Farley, J. M., Hackler, C. L., and Rilee, E. W., Jr., *Ceramic Bulletin*, Vol. 52, 1973, pp. 882–884.

[5] Snedecor, G. W. and Cochran, W. G., *Statistical Methods*, 6th ed., Iowa State University Press, Ames, 1967, pp. 86–88.

[6] Sokal, R. R. and Rohlf, F. J., *Biometry*; W. H. Freeman, San Francisco, 1969, pp. 716–721, and Chapter 5.

[7] Kateman, G. and Pijpers, F. W. *Quality Control in Analytical Chemistry*, Wiley-Interscience, New York, 1981.

[8] Dixon, W. J. and Massey, F. J., Jr., *Introduction to Statistical Analysis*, McGraw-Hill, New York, 3rd ed., 1969, Chapter 3.

[9] Kratochvil, B. and Taylor, J. K., *Analytical Chemistry*, Vol. 53, 1981, pp. 924A–938A.

[10] Harris, W. E. and Kratochvil, B. G., *Introduction to Chemical Analysis*, Saunders, Philadelphia, 1981, pp. 565–567.

[11] Laitinen, H. A. and Harris, W. E., *Chemical Analysis*, 2nd ed., McGraw-Hill, New York, 1975, pp. 576–578.

John K. Taylor[1]

Principles of Calibration

REFERENCE: Taylor, J. K., **"Principles of Calibration,"** *Sampling and Calibration for Atmospheric Measurements, ASTM STP 957,* J. K. Taylor, Ed., American Society for Testing and Materials, Philadelphia, 1987, pp. 14–18.

ABSTRACT: Measurement consists of comparison of samples of unknown composition with standards of known composition or with scales calibrated with respect to such standards. The standards used must simulate the unknowns with respect to matrix and level of analyte if the comparisons are to be valid. This paper reviews the fundamental aspects of calibration, describes various approaches that may be utilized, and considers the sources of error in the calibration process. The evaluation of calibration uncertainty and ways to minimize it are discussed. A general discussion of linear relationships as applied to calibration is presented.

KEY WORDS: calibration, chemical analysis, intercalibration, linear relationships, measurements, quality assurance, standardization, air quality, sampling, atmospheric measurements

The concept of calibration is very broad. As a noun, the word means a set of graduations marked to indicate values, such as the marking on a thermometer. As a verb, it means the comparison of a measurement standard or instrument with another standard or instrument to report, or eliminate by adjustment, any variation (deviation) in the accuracy of the item being compared. In chemical measurements, it refers to the process by which the response of a measurement system is related to the concentration or amount of analyte of interest, and hence often consists of the evaluation of an analog response function.

Standardization is a term related to calibration. As used by chemists, it means the establishment of the value of a potential chemical standard (a titrant, for example) usually with respect to a standard of known composition.

Regardless of the exact definition, the concept of calibration is basic to all of measurement. Measurement is essentially a comparison process in which an unknown whose value is desired is compared with a standard that is known [1]. The comparison may be direct, as in weighing with an equal-arm balance, or indirect using a previously calibrated instrument or scale. Even in the former case, the items compared are seldom identical so that some kind of calibrated instrument or scale is involved in evaluating the magnitude of any observed difference. Obviously, its calibration must be accurate if accurate measurement data are to be achieved.

What Needs to be Calibrated

The only purpose of calibration is to eliminate or minimize bias in a measurement process. Chemical measurements are made using a system that includes sampling and measurement processes. [2]. All aspects of the measurement process need to be calibrated. Likewise, all

---

[1] Coordinator for Chemical Measurement Assurance and Voluntary Standardization, Center for Analytical Chemistry, National Bureau of Standards, Gaithersburg, MD 20899.

aspects of the sampling process that involve measurement in any way (sieving, for example) and even the locating of sample sites should be calibrated. Ancillary data on such matters as temperature, pressure, humidity, particle size, volumetric capacity, mass, and flow rate may be needed, as well, requiring accurately calibrated instrumentation for their measurement. Accordingly, any of the instruments, standards, and methods used for these purposes must be calibrated to assure that their accuracy is within acceptable limits.

Requirements for Calibration

The prime requirement for calibration is the availability and use of appropriate and accurate standards. Minimally, they should have a high degree of similarity and, ideally, they should be identical to the object compared. In physical measurement, this is readily achievable but it is often difficult and sometimes approaches the impossible in chemical measurements. Even the effects of small deviations from matrix match and analyte concentration level may need to be considered and evaluated on the basis of theoretical or experimental evidence or both. Otherwise, accurate measurement data may be unachievable. Equally important is the accuracy of the standards used for calibration purposes. Chemical standards ordinarily are prepared by quantitatively combining constituents of known purity, but the latter cannot be automatically assumed. The purity of the analyte of interest needs to be known to the accuracy requirement of its measurement. When trace levels of analyte are of concern, even minute amounts of it in a diluent matrix can cause large calibration errors unless accurately measured and appropriate corrections are made for its presence. An analyst preparing his own standards should have the capacity of evaluating them. If not, there is an even stronger requirement to use only starting materials or standards prepared by suppliers of the highest reliability. In such cases, analysts should require proof of such reliability.

Similarly, the stability of standards, however and by whomever prepared, is a prime requirement. Every standard should have an assigned expiration date indicating its stable life expectancy and should not be used beyond such a date.

Frequency of Calibration

An important aspect of calibration is the determination of the maximum period between successive recalibrations [3]. Two basic and opposing considerations are involved: (1) the risk of being out of tolerance at any time of use, and (2) the cost in time and effort of calibration. The former consideration should be the major one because of the dilemma of what to do with data obtained during the interval between the last known in-calibration and first known out-of-calibration periods. However, an over-conservative approach could be prohibitively expensive. A realistic calibration schedule should reduce the risk of the former without undue cost and disruption to work schedules. The factors that need to be considered in a realistic calibration schedule include:

1. Accuracy requirement for the measurement data.
2. Level of risk involved.
3. Experience of the laboratory in use of the equipment or methodology.
4. Experience of the measurement community.
5. Manufacturer's recommendations.
6. External requirements for acceptability of data.
7. Cost of calibration.

An initial choice of calibration interval may be made on the basis of previous knowledge or intuition. Based on experience, the interval could be extended if the methodology is always within tolerance at each recalibration or should be decreased if significant out-of-tolerance is observed. Control charts [4] may be used to monitor the change of the measured value of a stable test item and to correlate the need to recalibrate. If a sufficiently large number of test instruments is involved, as in a monitoring program, a statistical study may be used to establish calibration intervals for critical items or procedures.

Mode of Calibration

Calibration of standards (perhaps standardization is a better term) is a static process. That is to say, calibration values should be independent of who does the calibration. Conversely, calibration of measurement systems is a dynamic process in that both the method of use and the user are involved. First, the system must be in statistical control since calibration's only function is to minimize bias. Then the question arises as to how much of the system to calibrate—the sensor, the analytical system, or the entire measurement system. The answer will depend on the sources of bias that need to be calibrated and the calibration interval required for each. The response of the sensor and the related instrumental analogue function will need frequent calibration and this is the easiest operation to perform. Often, only the introduction of a known gas mixture or test solution, or a series of such, is all that is required for this purpose. If judgment of the analyst is involved in a measurement process, a separate calibration could be required for the operation of the system by each analyst who will use it.

The entire measurement process should be calibrated for each class of test samples, perhaps less frequently, but never ignored. Calibration materials closely resembling the test samples will be needed for this purpose, but may be difficult or impossible to prepare or obtain. The analyst may be forced to use surrogates added to the test sample, or the method of standard additions of analyte, to calibrate the measurement system. Because artificially added analyte may not necessarily respond in the same manner as a naturally occurring analyte, the previously mentioned approaches may need to be validated.

The analyst must bear the responsibility for the adequacy of all calibrations related to his measurements. Methods of calibration, especially for physical test equipment, are the subjects of a number of American Society for Testing and Materials (ASTM) standards [5] and these should be used whenever applicable. Each chemical method should contain explicit procedures for its calibration but these may be limited to the measurement step or even to the sensor response. The calibration of the entire measurement system will require the use of appropriate standards as described earlier. The analyst must bear the full responsibility for their appropriateness.

Linear Relationships

Many chemical measurements do not involve direct comparisons with standards but rather indirect and intermittent comparisons in which the primary use of the standards is to establish an analytical response function that is used in subsequent analytical measurements. A linear function is typically appropriate, or instrument output may be linearized to achieve this condition. In such cases, the uncertainty of fit of the calibration function is a matter of concern [6].

In a typical chemical calibration situation, a series of calibration standards, considered to have insignificant compositional uncertainty, is measured one or more times and the response is plotted, after which a straight line is graphically drawn or fitted by the method of least-squares. The latter procedure is preferable, in that it provides an unequivocal estimate of the uncertainty of fit. Thus, the standard deviation of the slope and intercept can be calculated as well as the confidence interval for the line as a whole, a point on the line, and a single Y corresponding to a new value of X [7]. The joint confidence interval for the slope and the intercept also can be calculated to assist in deciding whether calibration lines obtained on various occasions are consistent [6].

It has been recommended that any mathematical fit of data be preceded by a graphical plot [7]. This will help to assure that a linear relationship is justified but also may be used to screen for outlying data points. Linear regression will try to accommodate all points by minimizing the sum of the squares of the residuals. The presence of outliers, ignored or given little weight when constructing a graphical plot, can be strongly influential in a least-squares fit and can compromise an otherwise excellent data set. There are no simple statistical criteria for rejection of outliers from graphical plots. The author uses the criterion of the "huge error," namely, a point that lies at least four times farther from the line than the average departure of all points is considered to be an outlier.

The uncertainty of the linear fit should be added in quadrature to that of the measurement process when overall uncertainty is being estimated. If calibration uncertainty is statistically significant and undesirably large, it can be reduced by multipoint or multimeasurement calibration or both.

In some calibration situations, such as ultra-trace analysis, the uncertainties of the standards used may not be insignificant. In such cases, the simple regression fit (x axis essentially errorless) is not justified [8,9]. The assistance of mathematical or statistical experts knowledgeable in handling such data may be required.

Calibration Uncertainties

Ideally, the calibration process is undertaken to eliminate deviations in the accuracy of measurement standards or instrument. However, this cannot be glibly assumed. In fact, as the limits of measurement are approached, the significance of the uncertainties of calibration may increase in a similar manner.

Overall uncertainty may be characterized according to the confidence warranted in the standards used and in the uncertainties of their use in the measurement process employing those standards. The uncertainty in the composition of chemical standards will depend on the degree of experimental realization of the calculated composition based on knowledge of the purity of constituents, on the accuracy of the blending process, and on considerations of stability. The reliability of the process for transferring the standard to the system calibrated is a further consideration. Both systematic and random sources of error are involved in all of these and will need to be minimized, consistent with the accuracy requirements of the data. Repetitive calibrations will decrease the random component of uncertainty but not any biases. As calibration uncertainty and measurement uncertainty approach each other, calibration can become a major activity, even in routine measurements.

If accurate standards could be prepared, identical or with insignificant dissimilarity to the unknowns, measurement would be simplified in that the analytical uncertainty would depend entirely on the precision of comparison. Only in rare cases is this possible. The complexity of most matrices (natural or manufactured) and the uncertainty (and virtual impossibility) of duplicating the physical and chemical interrelationships of the components of the matrix impedes the synthesis of equivalent standards. The sheer number of standards that would be required makes such an approach impractical even if it were possible to achieve. Accordingly, chemical calibration is an approximation, at best. The analytical chemist must be constantly aware of the possibility of bias introduced by the nature of the standards used that may be the major source of bias in the analytical data. Appropriate reference materials should be used to evaluate this and other aspects of the measurement process [10].

Intercalibration

Intercalibration may be defined as the process, procedures, and activities used to ensure that several laboratories engaged in a monitoring program can produce compatible data [11]. Such laboratories should have implemented a reliable and consistent calibration program and achieved statistical control before any intercalibration program can be meaningful. Intercalibration is best evaluated by systematic use of standard reference materials (SRMs) or similar substances. The reference materials should have a high degree of comparability with the test samples of the monitoring program, since the confidence in the measured values of the latter will be inferred from that of the former.

While at least one test level is necessary, this may not be sufficient in that it can only demonstrate intercalibration at that level. A minimum of three test levels (five is desirable) spanning the measurement range is necessary to evaluate the performance and to fully intercalibrate a measurement system [12].

Conclusion

By the very nature of measurement, calibration is one of its most critical steps. While the uncertainty of measurement can be worse than that of calibration, it can never be better. When statistical control has been verified, calibration uncertainties can be the major source of discrepancies between laboratories. Thus, biased measurement data can result when using unbiased methodology, because of calibration bias. Laboratories are urged to give more attention to this important operation and to critically evaluate it in every measurement situation. The consistent use of appropriate reference materials together with control charts can assure laboratories that unacceptable biases are not present in their measurements [13].

References

[1] Taylor, J. K., *Principles of Quality Assurance of Chemical Measurements*, NBSIR 85-3105, National Bureau of Standards, Gaithersburg, MD, 1985.
[2] Taylor, J. K. in *Quality Assurance for Environmental Measurements*, J. K. Taylor and T. W. Stanley, Eds., *ASTM STP 867*, American Society for Testing and Materials, Philadelphia, 1985, pp. 5–11.
[3] ILAC Task Force E, "Guidelines for the Determination of Recalibration Intervals of Measuring Equipment Used in Testing Laboratories," OIML International Document, No. 10, International Organization for Legal Metrology, Paris, France, 1984.
[4] *ASTM Manual on Presentation of Data and Control Chart Analysis, ASTM STP 15D*, American Society for Testing and Materials, Philadelphia, 1976.
[5] *Annual ASTM Standards, Index-Subject Index; Numerical List*, American Society for Testing and Materials, Philadelphia.
[6] Hunter, S. J., *Journal*, Association of Official Analytical Chemists, Vol. 64, 1981, pp. 574–583.
[7] Natrella, M. G., *Experimental Statistics*, NBS Handbook 91, No. 003-003-00135-0, Superintendent of Documents, U. S. Government Printing Office, Washington, DC, 1963.
[8] Mandel, J., *Journal of Quality Technology*, Vol. 16, 1984, pp. 1–13.
[9] Bartlett, M. S., *Biometrics*, Vol. 5, 1949, pp. 207–212.
[10] Taylor, J. K., *Journal of Testing and Evaluation*, Vol. 11, 1983, pp. 355–357.
[11] Taylor, J. K., "Quality Assurance for a Measurement Program," in Environmental Sampling for Environmental Waste, ACS Symposium Series 267, American Chemical Society, Washington, DC, 1985.
[12] Taylor, J. K., "Quality Assurance of Chemical Measurements," *Thalassia Jugoslavia*, Vol. 16, 1980, pp. 111–124.
[13] Taylor, J. K., *Handbook for SRM Users*, NBS SP260-100, National Bureau of Standards, Gaithersburg, MD, 1986.

Indoor Air—General

Hal Levin[1]

Overview of Indoor Air Quality Sampling and Analysis

REFERENCE: Levin, H., **"Overview of Indoor Air Quality Sampling and Analysis,"** *Sampling and Calibration for Atmospheric Measurements, ASTM STP 957*, J. K. Taylor, Ed., American Society for Testing and Materials, Philadelphia, 1987, pp. 21–34.

ABSTRACT: During recent years, sampling and analysis of atmospheres for indoor air quality in non-industrial environments have increased substantially. Enhanced understanding of indoor air pollution has resulted. Increased recognition of the complexity of indoor environments has led to even more sampling and analysis. Many methods used for indoor air are similar to those employed in workplace and ambient air. However, sampling conditions are often difficult; a very large number of pollutants are usually present, although at low concentrations; some pollutants of concern are at extremely low concentrations; and complex mixtures can create interferences.

Instrumentation requirements differ from those of ambient or industrial applications. Passive monitors and other unobtrusive monitoring equipment have received much attention. Sampling and analysis for biological aerosols will probably increase during the next few years. Other environmental factors are often critical and must also be characterized or measured. Social science and health investigatory methods are often employed in conjunction with air monitoring. Further research and development are needed for standardization in several areas including the sampling and analysis of organic compounds at very low concentrations, the general acceptance of standardized monitoring protocols, and guidelines for the consistent reporting and interpretation of data.

KEY WORDS: air quality, calibration, sampling, atmospheric measurements, indoor air quality, indoor air pollution, analysis, air quality monitoring, sampling instruments, air pollutants, indoor air pollutants, monitoring

Indoor air quality (IAQ) monitoring activities have increased substantially during the past eight years. The four principal types of monitoring are (1) investigations of "building sickness;" (2) surveys or studies for research purposes; (3) monitoring for occupancy or re-occupancy after contamination, and pre-occupancy evaluations of new buildings; and (4) compliance monitoring. Most IAQ sampling and analysis to date has been of the first two types. However, due to increasing awareness of the importance of IAQ, the latter three types are likely to be emphasized more frequently in the future.

Under its health hazard evaluation program, The National Institute of Occupational Safety and Health (NIOSH) has conducted a large number of investigations in offices, schools, and other public buildings in recent years. In Table 1, a summary of 203 investigations conducted between 1978 and 1983 indicates that the frequency of investigations increased significantly during that time [1].

The type of problems investigated by NIOSH are dominated by inadequate ventilation

[1] Research specialist, Center for Environmental Design Research, University of California, Berkeley, CA 94720.

TABLE 1—*Completed NIOSH investigations of indoor air quality by year (through December 1983)* [1].

Year	Number Completed	%
Pre-1978	6	3.0
1978	9	4.4
1979	12	5.9
1980	28	13.8
1981	80	39.4
1982	44	21.7
1983	24	11.8
Total	203	

(48.3%) with inside and outside contamination comprising an additional 17.7 and 10.3%, respectively. Other types of problems include humidty, building fabric contamination, hypersensitivity pneumonitis, and cigarette smoking, each representing >1% of the remaining investigations.

Ken Sexton (former director of the State of California Indoor Air Quality Program) reports that many laboratories surveyed by his agency are conducting indoor air pollution investigations that are neither for legal enforcement (compliance) nor for purely research purposes. These include investigations related to occupant complaints known or suspected to be of indoor air quality origin and owners' or tenants' desires to monitor buildings prior to occupancy [2].

While many IAQ studies are conducted for the purpose of determining the source of discomfort or health problems—building sickness—reported by building occupants, many other studies are conducted for research purposes. These include characterizing air quality in specific types of indoor environments or characterizing the distribution of specific pollutants in various building occupancy or construction types. Much research has focused on the emission rates or behavior of pollutants from specific building materials, equipment, appliances, consumer products, or furnishings [3].

Some indoor air monitoring has been for determination of indoor air quality after known contamination, such as incidents in office buildings where transformer explosions have resulted in widespread distribution of polychlorinated biphenyls throughout a structure. Other indoor air monitoring has involved preventative monitoring efforts intended to characterize IAQ prior to new building occupancy as well as in remedial work for problem buildings [2,3].

The symptoms of building sickness include eye, nose, and throat irritation; sensation of dry mucous membranes and skin; erythema (reddening of the skin); mental fatigue; headaches; high frequency of airway infections and cough; nausea; and dizziness [4,5]. Reports of building sickness have intensified efforts to understand its causes. These efforts have included air quality monitoring research investigations as well many epidemiologic investigations [5–7].

Nagda and Rector have made a distinction between research and investigation methods [8]. It is necessary to distinguish between monitoring methods utilized for building investigations, methods for enforcement of legal limits (compliance), and methods intended for research purposes. Monitoring used for building investigations differs from compliance monitoring since it often focuses on low levels of a large number of substances. It usually includes investigations of a building's mechanical equipment and the operating protocols and maintenance procedures employed in conjunction with this equipment. Some investigators also examine thermal parameters, humidity, and other factors that might contribute to the building sickness syndrome. Occasionally, health questionnaires, epidemiologic surveys, or clinical evaluations of occupants are also employed.

Standard methods and protocols for building air quality investigations do not yet exist. In general, equipment, methods, and protocols derived from ambient air quality monitoring and industrial hygiene are employed, frequently by individuals trained in those disciplines. While the failure of many investigators to determine the causes or sources of the complaints and health effects may be partially due to the application of methods developed for other purposes, the absence of guidelines or protocols to assist investigators in designing monitoring programs may also contribute to the lack of success. The American Society for Testing and Materials (ASTM) has recently established Subcommittee D22.05 on Indoor Air to develop standard methods and practices applicable to indoor air quality surveys and investigations.

In many indoor air quality investigations, specific pollutants are not found present in sufficiently large concentrations to explain the complaints and symptoms reported [5]. Some investigators hypothesize that the synergistic effects of low concentrations of large numbers of substances are involved in the etiology of the health effects [9]. No consensus has developed around the mechanism of action, the nature of the chemical mixture responsible, or measures other than improved ventilation to ameliorate the problem. This may well be due to measurement conditions that differ from normal use conditions. Most investigators recommend ventilation system improvements or operational changes.

It is necessary to develop protocols for assessing ventilation as part of indoor pollution monitoring work. Mechanical ventilation systems and natural ventilation found in most residences, offices, schools, and other non-industrial indoor environments usually differ substantially from industrial ventilation. There is no existing set of guidelines or methods widely accepted or used for their thorough investigation.

Normal "air balance" procedures for new buildings are intended to adjust the distribution of air among supply and return registers and to set supply and exhaust fan volumes to conform to design specifications. They do not, in fact, serve to quantify or evaluate ventilation. The actual performance of the ventilation system with respect to pollutant removal, adequacy of mixing within the conditioned spaces, variations from room to room, etc., are generally not evaluated. It is assumed that these are addressed adequately during system design. Experience does not validate this assumption [10]. Investigations conducted by NIOSH indicate that ventilation system problems are the most frequent causes of indoor pollution complaints [1].

In many cases of asbestos monitoring, mechanical ventilation systems are not operated during sampling. This often occurs because monitoring takes place on weekends or during school vacations and holidays to avoid disruption of school activities. Even when ventilation systems are operated, ventilation rates are normally not measured or estimated, nor are they reported. This raises questions regarding the validity of the monitoring data in assessing airborne asbestos levels.

While new methods and protocols are necessary for ventilation system evaluation and monitoring, there are large numbers of investigations, many of them reported in the literature, from which to draw information for the formulation of guidelines for such procedures. There are many researchers and professionals whose experience can contribute to an effective standards development process. And there are many new instruments and techniques developed explicitly for indoor air quality work that can provide alternatives to cumbersome, costly, or inappropriate ones developed for ambient or industrial environmental monitoring [8].

Important Indoor Air Pollutants

Asbestos and Other Fibrous Aerosols

Government and private asbestos hazard abatement activities have increased significantly in recent years. The U.S. Environmental Protection Agency (EPA) has published several editions of guidance documents for asbestos control in buildings [11].

A standard for asbestos monitoring, ASTM Test Method for Airborne Asbestos Concentration in Workplace Atmosphere (D 4240–83), uses phase contrast microscopy (PCM) analysis. There is a recognized need for the use of other analytical methods including polarized light microscopy (PLM) and transmission electron microscopy (TEM) for asbestos hazard assessment and control program evaluation. Guidelines for the design of ventilation monitoring and reporting protocols are needed to accompany indoor air quality asbestos monitoring.

Biologic Aerosols

Increasing attention to biologic components of indoor air has resulted from the work done to identify *Legionella p*. As a result, investigations have shown that airborne concentrations of viable organisms frequently correlate with physiologic responses and complaints. Symptoms including pulmonary manifestations, muscle aches, chills, fever, headache, and fatigue have been attributed to biologic agents. Attack rates in affected office buildings have been from 1 to over 50%. Disease has been attributed to thermophilic actinomycetes, non-pathogenic amoebae, fungi, and *Flavobacterium* spp. or their endotoxins [6]. Some bulk materials sampling may accompany air sampling routinely in future building investigations, research, and compliance monitoring. Increased attention to collection and analysis of such bulk samples along with air sampling methods and protocols for biologic aerosols will be developed and utilized more during the next few years.

Carbon Monoxide

Sources of carbon monoxide (CO) found indoors include tobacco smoke, gas and fossil fueled appliances, and adjacent garages and roadways. Monitoring is more frequent in residential than other indoor environments, although offices, schools, and other non-residential buildings are subject to contamination. Direct reading portable analyzers and larger real-time instantaneous instruments are available [12].

Carbon Dioxide

Direct reading portable analyzers as well as stationary and portable sampling devices can be utilized for carbon dioxide (CO_2) monitoring. Carbon dioxide measurements can provide indirect measures of ventilation rates in densely-occupied structures of known area and occupancy. Such measurements can be used for preliminary evaluations in buildings, both for compliance and for building investigation procedures. Work to date indicates that CO_2 is an excellent indicator of the relationship between occupancy and ventilation. All-outside-air ventilation can reduce by half the airborne CO_2 level compared with normal building ventilation operations [13]. It has also been reported that levels above 1000 to 1200 ppm are frequently associated with complaints related to building sickness symptoms [14]. Thus, measures of this sort can result in identification of the need for building ventilation investigations and corrective action prior to extensive chemical sampling and analysis that may be substantially more costly than ventilation measurements. Tracer gases and other means can be used to validate CO_2 findings, if necessary, for research or compliance purposes.

Formaldehyde

Formaldehyde is probably the best known indoor air pollutant. Its fame is due to its widespread use in building materials, especially plywood, panelling, and particle board. It is used as an adhesive or binder in these and other wood products as well as in insulations and in a host of other building products, furnishings, and consumer items. Indoor levels of 0.1 ppm are considered problematic and often are exceeded in new homes with low air exchange [15].

These levels are often associated with serious health problems and complaints of discomfort. The potential carcinogenicity of formaldehyde has been identified. The Interagency Regulatory Liason Group (IRLG) in its report to the Consumer Products Safety Commission, expressed concern about formaldehyde in combination with other substances found indoors. In particular, the IRLG cited chlorine, present indoors from vapors of hot water, that may combine with formaldehyde to form phosgene gas [16]. Formaldehyde is monitored with active and passive devices, and its measurement is almost routine in indoor pollution investigations [15,17–19].

Information on the sources and source strengths of formaldehyde emissions has been developed by several researchers including Matthews et al., Pickrell, and Meyer. Industry recently implemented an extensive material testing program, and on 11 February 1985, the Department of Housing and Urban Development (HUD) set formaldehyde emission standards for plywood and particleboard used in mobile homes. This is the first mandatory building material emission standard worldwide.

While such information has already lead to significant improvements in products and their uses, existing sources now in place will continue to contribute formaldehyde to indoor air for some time to come. Furthermore, many formaldehyde-containing building products, furnishings, and consumer products continue to be used indoors [15].

Inhalable Particulate Matter

Smoking, other combustion, and re-entrained dust appear to be important sources of inhalable particulate matter (IP) indoors. Indoor and outdoor concentrations and compositions differ significantly. The importance of IP exposures is accepted, but the measurement of indoor concentrations has resulted in increased emphasis on size distribution. Chemical adsorption on particulate surfaces is receiving additional attention as the organic chemical composition of indoor air becomes better documented. Although sampling instruments are available, they are quite noisy and may be unacceptable in many indoor air monitoring situations.

Metals and other inorganic particulate constituents (including lead, mercury, arsenic, and other heavy metals found in indoor air) are usually constituents of particulate matter (discussed later in this paper).

Nitrogen Dioxide

Nitrogen dioxide (NO_2) produced by combustion can be present in levels exceeding annual National Ambient Air Quality Standards (NAAQS). Many current research projects are monitoring NO_2, and the results will determine the extent of further work required. Work already completed indicates that NO_2 monitoring will continue to be an important component of indoor air quality work. Passive samplers and real-time instantaneous reading instruments are available.

Ozone

Office copiers and air cleaners are considered important sources of ozone indoors. Measured average levels of ozone in fully mixed indoor air may be low due to the reactivity of ozone and the normal abundance of organics. Levels at the operator breathing level near the ozone sources may be high and require special sampling protocols to provide the desired level of protection from excessive exposure. Care is also required in sampling when results will be the basis for comparing levels to other data.

Pesticides and Other Semi-Volatile Organics

Contamination of residential and non-residential environments by pesticides and other semi-volatile organics is not uncommon. Monitoring where severe contamination exists is often critical to decisions regarding occupancy of affected buildings [10,20–22]. Protocols for monitoring environmental conditions, for determining the extent of sampling, for determination and verification of ventilation system operation, and for assessing other relevant factors are necessary to provide the information used in making important personal, institutional, and public policy or regulatory decisions. Such protocols are lacking currently with the result that uncertainty and conflict can occur.

Polyaromatic Hydrocarbons (PAHs) and Other Organic Particulate Constituents

Polyaromatic hydrocarbons (PAHs) are complex organic substances including known and suspected carcinogens. Benzo-a-pyrene (BaP) is often measured to indicate PAH concentrations although the relationship is not well-defined. They are produced indoors by incomplete organic combustion including tobacco smoking, wood-burning, and cooking. They occur in vapor and condensed forms. Usually, only the particulate (condensed) form is measured [17].

Radon and Radon Progeny

Radon is an indoor pollutant of increasing concern as localized investigations in many parts of the United States uncover high indoor air levels. Health effects data based on mineworker exposure and epidemiologic studies provide reasonable estimates of health risks. Monitoring and remedial activities involve ventilation considerations. Passive and active monitors are available.

Sulfur Dioxide

Kerosene heaters appear to be the major source of sulfur dioxide (SO_2) indoors. Due to the extensive use of these devices for residential heating, monitoring is likely to continue.

Volatile Organics

Current EPA-sponsored research focused on volatile organic compounds (VOCs) indicates the number of compounds usually present is higher than previously believed. Furthermore, indoor-outdoor ratios are consistently greater than unity [23]. There is general agreement that VOCs are important indoor pollutants and that they present complex barriers to effective, affordable monitoring. State-of-the-art methods have not been the subject of standards that could assist the increasing number of investigators sampling for organic compounds.

Complex Mixtures

Tobacco smoke, exhaled human breath, cooking by-products, gas-combustion, and other indoor air constituents are composed of complex mixtures often including varieties of organic and inorganic vapors as well as particulates. These mixtures change over time and through space after generation as they mix with room air, combine chemically, and undergo physical changes. Source monitoring provides the basis for understanding their environmental fate. As more sophisticated control measures are considered, deeper understanding of complex mixtures will be required.

Much of the literature and research on indoor pollution focuses on tobacco smoke as an important source of indoor pollution. Tobacco smoke is considered to pose serious health threats to "innocent parties," that is, nonsmokers. As a result of both the health and irritant properties of tobacco smoke as well as increased office employment and reduced ventilation in office environments, a great deal of attention has been focused on passive or "involuntary" smoking. Legislation at the local level has been enacted or considered in a growing number of communities, most notably in San Francisco and neighboring municipalities in the Bay Area. As a result, employers are seeking ways to protect nonsmokers from the cigarette smoke of others and to discourage smoking in the work environment.

Characterization of cigarette smoke has been done for purposes of clinical and laboratory toxicity studies and epidemiology. However, little work has been done on the fate of the constituents of tobacco smoke in the indoor environment. It is likely that more such monitoring will be done in the coming years as regulatory and technical developments focus increasing attention on tobacco smoke as an indoor air pollutant and health hazard.

Efforts to identify effective means for controlling tobacco smoke concentrations have relied on measurements of specific components, usually particulates and carbon monoxide, occasionally formaldehyde and other organics, and, for special purposes, certain other constituents. However, insufficient study has been devoted to the measurement of tobacco smoke components at spatial or temporal distances sufficiently removed from the source to evaluate the effectiveness of various control measures. Yet ventilation standards are being developed that distinguish between smoking and nonsmoking occupancy in determining outside air requirements varying by factors from five to many times that in some recent risk analyses [24–26].

The environmental degradation products of many substances present indoors pose very serious health hazards. Included, for example, are the dioxins and furans that are formed when PCBs or PCPs are exposed to high temperatures. Since these chlorinated compounds are often contaminated with dioxins and furans during the manufacturing process, their hazardous by-products are present at low concentrations in most residential and office environments [22]. Emerging knowledge on the occurrence and hazards of these substances has already led to increased efforts to monitor them in instances such as the transformer explosions at Binghamton, NY, and San Francisco [20].

Non-Pollutant Measurements and Data

Indoor air quality monitoring is of little or no practical value where information on the environmental conditions, building operating conditions, building use patterns, and other factors affecting source strength and dilution are not reported. It is recommended that general background data on environmental conditions be reported with all indoor air quality monitoring, including research, compliance, or problem building investigation. The factors listed in this section are suggested as parameters for which standards be developed or adapted (where applicable standards already exists, as they do in many instances).

Thermal Environment

The thermal environment consists of the air temperature, the various surface temperatures within occupied spaces, and the temperatures of building materials that comprise heat sources or sinks. The significance of the thermal environment for indoor air quality evaluation is generally acknowledged [5,15]. Yet, monitoring of the thermal environment is not always reported in conjunction with indoor pollution monitoring. Furthermore, the thermal complexity of the indoor environment presents difficulties that must be considered in the design of monitoring programs. Air temperatures vary vertically within spaces, from air to surfaces, and within materials. These differences can have important consequences for both the thermal comfort of occupants as well as for the vapor pressures of substances being emitted into the air. Protocols for the monitoring of thermal parameters should accompany standards for measurement and reporting.

Relative Humidity

The moisture content of air significantly affects the release rate of many pollutants, the concentrations in air, and the potential to affect biological organisms and physical systems. A large number of devices exist that can economically and accurately monitor indoor air humidity within acceptable limits of cost.

Air Movement

Air currents within buildings are extremely complex and subject to the influence of a large number of factors. These include mechanical system functional characteristics, building plan, location and configuration of furnishings, activities, openings between interior spaces, openings in the exterior envelope, temperature differences between supply air and building air, and other factors.

Development of useful air sampling strategies and interpretation of results from monitoring require data on air movements within sampled spaces. Such data is necessary but difficult to acquire. A multiple tracer gas system may be developed to accomplish this. Further work is required. The following are some of the parameters that should be characterized:

1. velocity,
2. direction,
3. spatial patterns, and
4. volumes at supply and return registers in area monitored.

Air Exchange Rates

There are a number of methods now in use for determining air exchange rates and they vary greatly in cost, applicability, and reliability. These methods range from the use of tracer gas (that is, SF_6) to the use of pressurizing systems (so called "blower door systems" used in smaller structures, particularly houses, in conjunction with energy-related research). In addition, there are methods used by mechanical engineers and air balance specialists to measure air volumes at registers and fans in building mechanical systems. Then, by computation, overall outside air supply volumes are derived. Standardization of techniques for acquiring and reporting data are essential to make results comparable among buildings and investigations. Among the factors that can be characterized are the following:

1. ventilation rate—expressed as the volume of air per unit of time supplied to a specified building volume or area or per unit of occupant load; and
2. air exchange rate expressed as the volume of outdoor air as a percentage of ventilation air or as a volume of air per unit of time (outside air intake, exhaust air, infiltration).

Wind

Outdoor wind conditions can affect indoor air quality in a number of ways including changes in local or building ventilation rates and surface temperatures. The important wind parameters include the following:

1. direction,
2. velocity, and
3. variations at the perimeter of building.

Building Characteristics

Different types of investigations generate different needs for data on building characteristics. Standardization in the use of terms and the measurement practices involved in building characterization will improve the comparability of results and assist in the development of more systematic assessments. The list of topics in Table 2 is not exhaustive but illustrates the categories of data often sought in research and building investigation activities.

Protocols

The diversity of disciplines involved in indoor air quality work, the rapidly increasing interest in the subject, and the complexities and difficulties of developing effective monitoring protocols create the need for guidelines regarding the development of protocols. These may divide into two classification systems: (1) type of investigation, and (2) suspected pollutants or sources.

The type of investigation will play a major role in determining the requirements of the protocol. Research, compliance, and building investigations each have separate requirements for protocols. Different pollutants or sources also determine protocol design. The following sections describe some considerations in the development of protocols and will form the basis for discussions on the incorporation of protocols into indoor air quality monitoring standards.

Additionally, where particular sources are suspected, or where monitoring is conducted for determining product emission rates, special protocols are required.

General Considerations

"The aim of sampling is to provide a miniature reproduction of the larger portion of the environment that is to be examined, but on a scale that will enable the sample to be manipulated in the laboratory" [27].

TABLE 2—*Important building characteristics in indoor air quality monitoring programs.*

A. Indoor/outdoor relationships
    1. Openings in building envelope: size, location, accessibility
    2. Nature of operational barriers (doors, windows, vents, screens, dampers, filters, fans, etc.)
    3. Timing and use characteristics of openings
    4. Control of windows, shades, blinds, curtains
    5. Monitoring
    6. Envelope permeability to light, heat, particles, gases
B. Occupancy
    1. Density and distribution of population
    2. Population characteristics
    3. Activities
    4. Schedules
C. Maintenance of building finishes, environment
    1. Activities: sweeping, dusting, vacuum, waxing
    2. Schedules: frequency
    3. Materials: type and quantities of chemicals
    4. Lighting replacement, window cleaning
D. Mechanical system
    1. Description: design, configuration, components
    2. Operations: timing, modes, control
    3. Maintenance: frequency, procedures, evaluation, verification
    4. Performance: measured air volumes, temperatures, humidity
    5. Filter renewal frequency
E. Indoor "climate"
    1. Thermal
    2. Moisture
    3. Air movement
    4. Ventilation rate (building air exchange with outside)
    5. Ventilation efficiency (air distribution effectiveness at the human breathing zone)
    6. Light: quantity, quality
    7. Noise: background, excursions (intensity and duration)
    8. Odors

Sampling in enclosed environments (industrial and non-industrial workplace, indoor non-workplace) differs from outdoor sampling in several aspects. Among the most important are the introduction of ventilation air through mechanical or natural means, the diversity of pollutant sources, the very large number of sources and contaminants of concern, and the imperfect mixing of indoor air resulting in special difficulties. Of these difficulties, representation of the larger indoor environment of which the sample is to be representative requires attention to temporal and spatial variations.

Even where indoor conditions are closely controlled by mechanical equipment, variations in meteorology and outdoor source strength can significantly affect distributions and concentrations of pollutants indoors. The most careful and thorough planning of monitoring activities will often be undermined by changes in occupant behavior (not smoking, different activity patterns, etc.) or climatic variations. Mixing can be less complete indoors than outdoors due to the limited volumes, large surface area variations, and the action of natural and mechanical ventilation systems.

Sampling in indoor air may present special problems due to the unacceptability of invasive sampling equipment and personnel. Noise, displacement of objects, restrictions on movement or activities, modifications to normal air handling protocols, etc. are often unacceptable aspects of indoor air sampling. The same problems often apply to the collection of data on other environmental parameters.

Operating conditions of the building cannot always be modified to provide the desired test conditions—worst case, typical case, best case, 100% outside air, minimal outside air supply, standard temperatures, or humidities. Thus, testing must often be done under conditions that might differ significantly from desired conditions.

The tendency for many pollutants to adsorb onto the very large surface areas presented in indoor environments results in dynamics that limit the ability to compute or infer concentrations under conditions other than those at which testing was performed. Furthermore, factors affecting pollutant concentrations prior to monitoring must be ascertained and, to the degree possible, controlled to assure steady state and, therefore, well-defined concentrations have been reached at the time of monitoring.

Reports of monitoring for several indoor pollutants show substantial diurnal, seasonal, and annual variations due to meteorological factors, building operational factors, and source strength factors [15].

Indoor Air Quality Research

Some monitoring will be done for the purpose of determining the efficacy of monitoring protocols, instruments, analytical techniques, or other aspects of indoor air quality. Other investigators are interested in estimating population exposure to indoor pollutants. This work includes that done for purposes of characterizing indoor pollutants in various building types and work done to assist in the development and validation of models used for predictive or policy purposes. Other work is conducted to estimate source strengths (product emission rates) or to determine the efficacy of various control measures.

Due to the purpose and use of research work, quality assurance and quality control measures must be rigorously applied. Furthermore, reports of results must be more comprehensive and detailed including data on related factors such as environmental conditions indoors and outdoors, concentrations of other pollutants, characteristics of the building being monitored, and other factors that may affect the interpretation or replicability of the results by other researchers or by policy-makers.

Compliance Monitoring

Monitoring conducted to determine compliance with legal limits or consensus standards must be done under standardized conditions or ones that can be referenced to standardized conditions to avoid false negatives or positives. Prevailing restrictions on funding of public agencies ordinarily involved in compliance monitoring require that methods be as economic as possible. The potential significance of the results on building owners, occupants, and users requires that the results be reliable and replicable. Quality assurance and quality control require thorough documentation.

Investigations of Problem Buildings

Investigations are conducted by professionals and technicians with a wide range of backgrounds, interests, and skills. Investigations are conducted under widely varying circumstances. Weather, occupancy, operation of building equipment, etc. that can have significant impacts on monitoring results can vary substantially from one building to another and from one day to another in a single building. The purpose of such investigations is to determine the source of dissatisfaction with indoor air quality or of health effects.

Since both the problems being investigated and the building environment can vary so widely from case to case, it will not be possible to develop universal protocols. However, there are guidelines that can be provided to assist in the design of protocols and assessment of monitoring results.

Due to the variations between buildings and the personnel involved in investigations, it is important that guidelines be provided to reduce useless or misleading monitoring. It is also important that variables (other than the concentrations of indoor air pollutants) that may contribute to the problem be considered systematically, preferably before air monitoring is conducted.

The experience of investigators who fail to identify the specific source of complaints indicates that there may either be synergistic effects of various pollutants and other environmental factors or that investigations may systematically fail to monitor effectively. One strong possibility emerging from the literature on investigations is that investigators arrive too long after episodic pollution that may originate in new materials, furnishings, or equipment. Particularly important may be volatile, low molecular weight organic compounds that are present in some materials. The long lag time between onset of symptoms leading to complaints and the reports of complaints or the initiation of investigations have hampered investigations in the past.

Pre-Occupancy Evaluation

General awareness of indoor air quality issues has resulted in the monitoring of some new buildings prior to initial occupancy. In such instances, manipulation of building ventilation and the environmental factors it controls allows for a variation in conditions to suit the particular building and situation. Post clean-up monitoring in contaminated buildings also allows for more complete control of environmental conditions. Buildings contaminated by asbestos constitute a major portion of this type of monitoring, although there are reported cases of clean-up activities or monitoring related to contamination by PCBs, pentachlorophenol, and other pesticides. There is very little published literature on this type of monitoring [10,20,21].

Materials Evaluations

There is a growing demand for data on the emission rates of chemicals from building materials, furnishings, and equipment. Private industry trade associations and public and private laboratories are conducting tests. A small number of test methods have gained favor. These include vacuum extraction, desiccator tests, head space tests, and environmental

TABLE 3—*Factors to consider in the selection of instruments for indoor air monitoring.*

1. Personal, portable, or stationary
2. Sensitivity/limits of detection
3. Cost of instrument
4. Calibration requirements and procedures
5. Accuracy within operating range
6. Power requirements, service time
7. Versatility for more than one pollutant
8. Integrated or direct reading
9. Recording capability
10. Obtrusiveness: noise, visibility, space requirements
11. Active or passive
12. Analyzer or collector
13. Availability
14. Travel considerations

chamber tests. Standard methods already exist for some of these tests. Others are completely without standardization. The American Society for Testing and Materials is making a valuable contribution by developing such standards.

Instruments

A variety of instruments have been developed in recent years that are applicable to indoor air quality work. Many of them have been developed explicitly for indoor pollution monitoring. Others were developed for other purposes but may be applicable.

Instruments available for indoor air quality monitoring were reviewed in 1982 by Sandia National Laboratories and by Geomet, which is currently preparing updates for their earlier report. A review of instruments including the topics listed in Table 3 will be useful as part of the general documentation that accompanies the completed standards compendium [8,28]. Some categories for review or analysis of instruments for use indoors are given in Table 3.

Analytical Methods and Considerations

Major innovations in analytical methods have been made, particularly for organic compounds and asbestos fibers. However, there are controversies as well as new developments concerning the analysis of these substances and others found in indoor air. It is important to the overall value of the standards that broad participation and state-of-the-art methods be considered and evaluated in the standards development process [8,27,29,30].

Quality Assurance and Quality Control (QA/QC)

Existing standards and methods commonly contain guidelines for quality assurance and quality control (QA/QC) procedures, criteria, and standards. This information can be modified to make it suitable to the indoor air quality monitoring activity at hand: research, compliance, or building investigation. While there may be differences in the rigor applied in each type of monitoring, the provision of standards will establish guidelines and the basis for adoption of more rigorous QA/QC practices where applicable [8,28,31,32].

References

[1] Melius, J., Wallingford, K., Keenlyside, R., and Carpenter, J., *Evaluating Office Environmental Problems*, Annals of the American Conference of Governmental Industrial Hygienists, Vol. 10, 1984, pp. 3–7.

[2] Sexton, K., *Journal*, Air Pollution Control Association, Vol. 35, No. 9, 1985, pp. 626–631.

[3] Levin, H. in *General Proceedings, Research and Design 85: Architectural Applications of Design and Technology Research*, American Institute of Architects, Washington, DC, 1985, pp. 173–179.

[4] Finnegan, M. J., Pickering, C. A. C., and Burge, P. S., *British Medical Journal*, Vol. 289, 1984, pp. 1573–1575.

[5] *Indoor Air Pollutants: Exposure and Health Effects*, World Health Organization, Copenhagen, 1982.

[6] Morey, P. R., Hodgson, M. J., Sorenson, W. G., Kullman, G. J., Rhodes, W. W., and Visvesvara, G. S., *Evaluating Office Environmental Problems*, Annals of the American Conference of Governmental Industrial Hygienists, Vol. 10, 1984, pp. 21–35.

[7] Turiel, I., *Indoor Air Quality and Human Health*, Stanford University Press, Stanford, CA, 1985.

[8] Nagda, N. and Rector, H., "Guidelines for Monitoring Indoor Air Quality," EPA 600/4-83-046, Geomet Technologies, Inc., Rockville, MD, 1982.

[9] Hollowell, C. and Miksch, R., *Bulletin*, New York Academy of Medicine, Vol. 57, No. 10, 1981, pp. 962–977.

[10] Levin, H. in *Proceedings*, 3rd International Conference on Indoor Air Quality and Climate, Stockholm, Swedish Council for Building Research, 20–24 Aug., 1984, Vol. 3, pp. 123–130.

[11] "Guidance for Controlling Asbestos-Containing Materials in Buildings," EPA 560/5-85-024, U.S. Environmental Protection Agency, Washington, DC, 1985.

[12] Nagda, N. and Koontz, M. D., *Journal*, Air Pollution Control Association, Vol. 35, 1985, pp. 134–137.

[13] Turiel, I., Hollowell, C. D., Miksch, R. R., Rudy, J. V., Young, R. A., and Coye, M. J., *Atmospheric Environment*, Vol. 17, No. 1, 1983, pp. 51–64.

[14] Rajhans, G. S., *Occupational Health in Ontario*, Vol. 4, No. 4, 1983, pp. 160–167.

[15] Meyer, B., *Indoor Air Quality*, Addison-Wesley, Reading, MA, 1983.

[16] "Final Report on Formaldehyde," submitted to the Consumer Products Safety Commission, Washington, DC, Interagency Regulatory Liason Group, Federal Panel on Formaldehyde, 1980.

[17] *Indoor Pollutants*, National Academy of Sciences Press, Washington, DC, 1981.

[18] Matthews, T. G., Daffron, C. R., Reed, T. J., and Gammage, R. B., "Modeling and Testing of Formaldehyde Emission Characteristics of Pressed-Wood Products, Report XII," Oak Ridge National Laboratory, Oak Ridge, TN, 1983.

[19] Pickrell, J. A., *Environmental Science and Technology*, Vol. 18, No. 9, 1984, pp. 682–686.

[20] Hahn, J., Tappen, D., Conners, M., and Levin, H., "Final Report on Environmental Monitoring at One Market Plaza for the Office of the Comptroller of the Currency," Cooper Engineers, Richmond, CA, 1984.

[21] Jurinski, N. in *Proceedings*, 3rd International Conference on Indoor Air Quality and Climate, Stockholm, Swedish Council for Building Research, 20–24 Aug., 1984, Vol. 4, pp. 51–56.

[22] MacLeod, K. E., "Sources of Emissions of Polychlorinated Biphenyls into the Ambient Atmosphere and Indoor Air," EPA 600/4-79-022, U. S. Environmental Protection Agency, Research Triangle Park, 1979.

[23] Wallace, L., "Organic Chemicals in Indoor Air: A Review of Recent Findings," presented at American Association for the Advancement of Science Annual Meeting, New York, 26 May, 1984.

[24] "Ventilation for Acceptable Indoor Air Quality," Standard 62-1981, American Society of Heating, Refrigeration and Air Conditioning Engineers, Atlanta, 1981.

[25] Repace, J. in *Proceedings*, 3rd International Conference on Indoor Air Quality and Climate, Stockholm, Swedish Council for Building Research, 20–25 Aug., 1984, Vol. 5, pp. 235–239.

[26] Weber, A., in *Proceedings*, 3rd International Conference on Indoor Air Quality and Climate, Stockholm, Swedish Council for Building Research, 20–24 Aug., 1984, Vol. 2, pp. 297–301.

[27] *Principles for Evaluating Chemicals in the Environment*, National Academy of Sciences, Washington, DC, 1975.

[28] *Occupational Exposure Sampling Strategy Manual*, U. S. Public Health Service, Washington, DC, 1977.

[29] "Compendium of Methods for the Determination of Toxic Organic Compounds in Ambient Air," EPA 600/4-84-041, U.S. Environmental Protection Agency, Research Triangle Park, 1984.

[30] *Estimating Human Exposure to Air Pollutants*, World Health Organization, Geneva, 1982.

[31] "Quality Assurance Handbook for Air Pollution Measurement Systems; Volume 1—Principles," EPA 600/9-76-005, U. S. Environmental Protection Agency, Research Triangle Park, 1976.

[32] Wadden, R. A. and Sheff, P. A., *Indoor Air Pollution: Characterization, Prediction, and Control*, Wiley, New York, 1983.

[33] *Selected Methods of Measuring Air Pollutants*, World Health Organization, Geneva, 1976.

Roy C. Fortmann,[1] *Niren L. Nagda,*[1] *and Michael D. Koontz*[1]

Indoor Air Quality Measurements

REFERENCE: Fortmann, R. C., Nagda, N. L., and Koontz, M. D., "**Indoor Air Quality Measurements,**" *Sampling and Calibration for Atmospheric Measurements, ASTM STP 957,* J. K. Taylor, Ed., American Society for Testing and Materials, Philadelphia, 1987, pp. 35–45.

ABSTRACT: Measurements of contaminants and other parameters in indoor environments present unique challenges to researchers because of the variety of building types to be monitored, the presence of occupants, the large number of potential contaminant sources, and the high degree of temporal and spatial variation of contaminant concentrations.

Selection of monitors and sampling methods for indoor air quality monitoring requires careful consideration of numerous factors. Considerations related to the design of the indoor monitoring program as well as factors related to monitor and sampler performance must be evaluated. Practical considerations related to sampling in occupied environments play an important role in the selection of measurement devices.

This paper discusses general concepts that should be considered in the selection of instrumentation for monitoring indoor air quality. Specific instrumentation and methods are not reviewed, but examples of instrumentation and monitoring approaches are provided for researchers unfamiliar with indoor air quality monitoring.

KEY WORDS: air quality, indoor air quality, air pollution, air pollution monitoring, analytical methods (air), calibration, sampling, atmospheric measurements

Measurements of air quality in indoor environments have gained increased importance in recent years with the recognition that concentrations of some contaminants indoors, where people spend 80 to 90% of their time, are higher than outdoors [1–3]. Consequently, measurement of indoor contaminant concentrations is essential to evaluate total human exposure. In addition to measurements of contaminants, other factors related to the characteristics of the building and its occupants need to be measured to understand the significance of the measured contaminant concentrations.

Many different pollutants are recognized to be important indoors. The list of contaminants that may need to be monitored include asbestos and other fibrous aerosols, biogenic aerosols, inhalable particles, carbon monoxide (CO), carbon dioxide (CO_2), nitrogen dioxide (NO_2), ozone (O_3), sulfur dioxide (SO_2), radon (Rn), radon progeny (RnP), formaldehyde (HCHO) and other aldehydes, volatile organic compounds (VOCs), pesticides and other semivolatile organics, and polynuclear aromatic hydrocarbons (PAH) and other organic constituents of particulate matter.

In addition to air contaminants, other factors need to be measured in indoor air quality (IAQ) monitoring programs to fully understand the significance of contaminant measurements. Important factors to be considered in IAQ studies include air exchange rates, building design and ventilation characteristics, indoor contaminant sources and sinks, air movement and mixing, temperature, relative humidity, and outdoor contaminant concentrations and meteorological conditions.

[1] Assistant director, Indoor Environment Division; director, Indoor Environment Division; and senior research scientist; respectively, GEOMET Technologies, Inc., Germantown, MD 20874.

A wide variety of measurement devices and methods are available with potential application to IAQ monitoring programs. However, the selection of such devices and methods for IAQ monitoring requires careful consideration of many factors. Some of these factors are discussed in this paper as they relate to design of IAQ monitoring programs and selection of measurement devices and methods. The discussion is supplemented with examples.

There are many research groups that have provided a significant contribution in this area; references to their work is included in the text of this paper. However, it should be noted that this paper is not intended to be a comprehensive review of instrumentation currently available for IAQ monitoring. Such a treatment would be a subject for a book or a monograph. A limited number of examples of monitors, samplers, and monitoring approaches based on recent studies conducted by authors are presented as a general introduction for those unfamiliar with IAQ monitoring.

IAQ Measurement Devices and Methods: Definitions

General categories of measurement methods or devices used for IAQ measurements can be defined as follows:

Monitors—Instruments that both collect and analyze the sample on a real-time basis. These devices are usually automated and include provisions for output of a digital or analog signal. The output may represent continuous measurements that parallel actual contaminant concentrations, semicontinuous instantaneous measurements, or integrated measurements for a preselected time period. Sample delivery to the sensor may be active or passive, but some power source is usually required for analysis.

Active Samplers—Contaminant collectors that require a power source with pump to deliver the air sample to a sensor or collector. Analysis of the contaminant concentration is performed subsequent to sample collection, usually in a laboratory.

Passive Samplers—Contaminant collectors that rely on diffusion for delivery of the sample to the collector; a power source is not required. Sample analysis is performed subsequent to collection.

Survey Instruments—Questionnaires and activity logs used to record pertinent information on factors related to IAQ. Information on factors such as building volume, age, and ventilation system; occupant density and activities; and indoor contaminant sources is often recorded with questionnaires. Factors that may change with time, such as occupant use of appliances, window and door openings, and other occupant activities, can be recorded with activity logs.

The distinction between active and passive methods of sample collection is particularly important for monitoring indoor environments. As will be discussed in the following sections, the availability of unobtrusive, low-cost passive monitors and samplers is essential to the successful performance of IAQ monitoring in some situations. A further distinction to be made is related to sampling mobility. Three classes of mobility can be defined as follows:

Personal—The monitor or collection device can be worn or carried by a person without substantially affecting normal activities.

Portable—The monitor or sampler can be easily moved from site to site, but not worn.

Stationary—The monitor or sampler must be operated at a fixed site.

Instrument Availability

A variety of monitors and samplers designed for ambient (outdoor) monitoring are available that are potentially applicable to indoor monitoring; to much more limited extent, some methods used for industrial workplace monitoring may be applied indoors. In recent years, a number of instruments and techniques have been developed or redesigned specifically for use in indoor environments. Both active and passive devices are available. Passive devices are particularly attractive for IAQ monitoring because of their ease of use, small size, and generally low cost. Personal and portable monitors are finding widespread use in IAQ monitoring programs because their use is generally more applicable indoors than stationary monitors.

A number of sources of information are available that describe monitors and samplers applicable to IAQ monitoring. Personal monitors that can be used for either personal or area monitoring indoors have been reviewed by Wallace and Ott [4]. Available methods for measurements of selected pollutants have been compiled in a report by the Sandia National Laboratories [5]. Hawthorne, Matthews, and Vo-Dinh [6] published a review of measurement techniques that includes a comprehensive reference list of relevant technical literature. A 1983 report by Nagda and Rector [7] provided a review of available monitoring instrumentation and methods. This source has been updated and expanded by Nagda, Rector, and Koontz [8] in a book that contains comprehensive appendixes that list available monitoring methods and instrumentation, performance and operational characteristics, manufacturers, costs, and relevant literature references.

Techniques for measuring air infiltration were reviewed at the 1980 Air Infiltration Center (AIC) conference by Harrje, Grot, and Grimsrud [9]; proceedings of subsequent AIC conferences provide a valuable source of information on current techniques. Information on measurement methods can also be found in various papers presented at two international conferences on indoor air quality [2,3]. Additional sources of information on measurement methods include publications by the American Society for Testing and Materials [10], the American Conference of Governmental Industrial Hygienists [11], and Linch [12]. However, the last three publications do not address the needs of indoor air quality monitoring specifically; applicability of methods in these volumes must be evaluated in terms of the measurement range, sensitivity, interferences, etc., required for IAQ monitoring. These factors are discussed in a following section of this paper.

The importance of survey questionnaires and activity logs as instruments to be used for data collection in IAQ monitoring has not been fully recognized. Therefore, information on their design is limited. Standard formats for survey questionnaires are not available and technical reports of IAQ studies rarely include examples of survey questionnaires. The design of survey instruments has been addressed by Koontz and Nagda [13] and Nagda, Rector, and Koontz [8]. Survey instruments will not be discussed in this paper due to space limitations, but their importance and inclusion in any IAQ monitoring program should be recognized.

Selection of Monitors and Methods for IAQ Monitoring

Program Design Considerations

In the design of a IAQ monitoring program, a systematic approach should be used to ensure that the results of the monitoring program adequately address the study objectives. Nagda, Rector, and Koontz [8] suggest that the preliminary research plan for an IAQ monitoring program should include the following steps: (1) develop monitoring objectives from the study objectives, (2) identify the range and extent of sampling requirements, (3) identify available instrumentation and methods for measurements, and (4) develop the initial design. The preliminary research plan is then used as the basis for development of a detailed monitoring program and final selection of the measurement methods. As indicated by inclusion of Step 3, the availability of monitors and samplers plays an important role in the design of the monitoring program.

The initial design of an IAQ monitoring program includes many considerations that affect the selection of monitors and samplers. Some of these design considerations are presented in Table 1. In any monitoring program, the study objectives are the major consideration in the design of the program and ultimately the selection of measurement devices. Examples of IAQ study objectives include the following:

1. To characterize contaminant concentrations in buildings with known contaminant sources; for example, unvented combustion appliances.
2. To establish the relationship between contaminant concentrations and building materials or occupant activities.
3. To test contaminant mitigation strategies.
4. To identify contaminants or other factors contributing to building-associated illness complaints.

In many cases, the study sponsor will suggest which contaminants and buildings are to be monitored as well as the desired outputs; for example, whether average or peak contaminant

TABLE 1—*Indoor air quality monitoring program design considerations that affect the choice of measurement methods and monitors.*

Study objectives and desired output
Quality assurance objectives
Pollutants to be monitored
Other parameters to be monitored
Sample size:
Number of buildings
Number of locations per building
Duration of monitoring/sampling period
Monitoring location (geographic)
Monitoring period (season)
Type of building(s):
Age
Size and design
Ventilation/environmental control system
Use
Construction materials
Accessibility for IAQ monitoring
Occupants:
Density and distribution
Characteristics
Activities
Ages
Willingness to cooperate
Contaminant sources:
Number
Type
Release mode (continuous or sporadic)
Contaminant concentrations, range, and variability

concentrations are to be measured. In other cases, these decisions will be left to the investigator and must be based on the study objectives and available resources for monitoring.

A comprehensive quality assurance (QA) plan that considers the 16 elements described in guidelines prepared by the U. S. Environmental Protection Agency [14] is an essential component of any IAQ monitoring program. The QA plan should specify the manner in which the monitoring program will attempt to achieve predetermined goals of data quality and include QA objectives, custody provisions, internal quality control checks, performance and system audits, and corrective actions. The QA objectives of accuracy, precision, and completeness of the data should be considered early in the design of the monitoring program and be used as criteria in the selection of monitors and methods. However, the final specification of project QA objectives will depend on the performance characteristics of the monitors and methods available for the project. Therefore, although QA objectives are considered as criteria in selection of measurement methods, the project QA objectives cannot be defined until all components of the measurement system have been selected.

An important consideration in the design of an IAQ monitoring program is the selection of other parameters to be measured in addition to contaminants. Measurements of indoor contaminant concentrations alone may be of little value if other parameters such as air exchange rates, temperature, relative humidity, or outdoor contaminant concentrations are not measured. A useful approach for determining what other factors to measure is to consider the terms of the mass balance equation. A simplified mass balance equation for a single compartment [1] states that:

$$\frac{dx}{dt} = \frac{VdC_{in}}{dt} = \begin{array}{l} \text{rate of change} \\ \text{in mass due to} \end{array}$$

$$\times \left[\begin{array}{l} \text{infiltration of} \\ \text{outdoor air} \end{array} + \begin{array}{l} \text{generation} \\ \text{indoors} \end{array} - \begin{array}{l} \text{exfiltration} \\ \text{of indoor air} \end{array} - \begin{array}{l} \text{indoor removal} \\ \text{of contaminants} \end{array} \right]$$

where V is the indoor volume and C_{in} is the indoor concentration.

The mass balance terms include outdoor concentration, air exchange rate, volume of the structure, source generation rate, removal and decay rates, filtration, and mixing factors. Measurement of parameters related to the terms of the mass balance equation strengthens data interpretation, improves the comparative value of the data, and allows projection of the results to other conditions.

Many of the design considerations listed in Table 1 have a major impact on the selection of monitoring instrumentation. Building size and design, for example, dictate the number of sampling locations required and, therefore, the number of monitors required to obtain representative measurements of air contaminants. In some cases, building size and occupant density and distribution will determine whether portable or stationary monitors are applicable to the monitoring situation. Building accessibility influences when samples are collected and the maximum or minimum duration for sample collection. Occupant density and age have a strong influence on selection and siting of measurement devices; the presence of small children, for example, is an important consideration in monitoring residential dwellings.

Monitor and Sampler Selection Factors

The choice of measurement devices for IAQ monitoring programs should be based on consideration of both the monitoring program design and factors related to the measurement devices. These factors include things such as instrument size, cost, calibration and maintenance requirements, and performance specifications. Table 2 presents a "generic" list of factors.

Failure to select the proper measurement device may result in serious logistical problems in the conduct of the monitoring program, high monitoring costs, or worse, monitoring results that are insufficient to meet the study objectives. The key to proper selection of monitors and methods for IAQ monitoring is to consider the selection factors listed in Table 2 as they relate to the monitoring program design considerations presented in Table 1.

Some examples of the relationship or interaction between design considerations and instrument selection factors are presented in Table 3. As an example, electrical wiring in older buildings may be inadequate for the requirements of stationary monitors with large pumps or transformers. Additionally, such monitors may have high heat output that affects air movement indoors at the sampling site. Numerous contaminant sources may require operation of instruments on ranges higher than normally used for ambient monitoring. The presence of building occupants may preclude use of noisy instruments for even short periods.

Common sense is the researcher's most valuable asset when selecting instrumentation for IAQ monitoring. Large stationary instruments may be attractive for obtaining very precise and accurate results, but maintenance, calibration, and labor costs may be excessive if they are used for short periods at many locations in large surveys. The presence of small children is a major consideration in selection and deployment of instruments because of children's propensity to tamper with dials and switches, particularly if accompanied by flashing lights.

Basic questions that the researcher should consider in the design of an IAQ monitoring program and selection of monitors and samplers should include, but not be limited to the following:

1. How will the indoor environment to be monitored impact on the performance of the monitor/sampler?
2. How will the monitor/sampler impact on the indoor environment being monitored?
3. How will the monitor/sampler impact the occupants and their routine activities?
4. How will the occupants and their activities impact on the IAQ measurements?

TABLE 2—*Factors for consideration in the selection of measurement methods and monitors.*

Mobility:	Interferents
Size	Selectivity
Weight	Recording capability
Operating specifications:	Type of output:
Power requirements	Continuous
Range	Semicontinuous
Sample flow rate	Integrated
Environmental operating range	Obtrusiveness:
Heat output	Noise
Cooling requirements	Space
Unattended operational period	Device security provisions
Minimum required sampling period	Labor requirements:
Exhaust requirements	Calibration
Performance specifications:	Operation
Accuracy	Maintenance
Precision	Analysis
Minimum detection limit	Costs:
Linearity	Purchase
Range	Operation
Zero and span drift	
Lag, rise, and fall times	

Monitoring Approaches

The final detailed design of an IAQ monitoring program is the result of careful consideration of the factors just described. A well-designed program must meet three general objectives:

1. output of data to meet study objectives,
2. implementation with available resources, and
3. logistical feasibility.

A variety of monitoring approaches or designs can be used to meet these objectives. A limited number of examples that illustrate monitoring approaches used for IAQ studies are described in this section.

Mobile Laboratories

Mobile laboratories equipped with real-time monitors have been used by various groups for IAQ monitoring. The EEB Mobile Laboratory operated by the Lawrence Berkeley Laboratory is a semitrailer instrumented for real-time monitoring of concentrations of CO, CO_2, SO_2, nitrogen oxides (NO_x), NO_2, O_3, indoor temperature and moisture, infiltration, and outdoor meteorological conditions [15]. Instrumentation is also included to collect integrated samples of Rn, HCHO, total aldehydes, VOCs, and inhalable particles.

The GEOMET Mobile Laboratory includes real-time monitors for NO_x, NO_2, CO, CO_2, O_3, SO_2, Rn, RnP, methane, total hydrocarbons, indoor temperature and moisture, outdoor meteorological parameters, and air exchange. Other equipment in the laboratory is used

TABLE 3—*Examples of the relationship between design considerations and measurement device selection factors.*

Program Design Considerations	Instrument Selection Factors	Relationship
Building age	power requirements	electrical systems may be inadequate in older buildings
Building size	sample flow rate	high-volume samplers may effectively "vacuum" the air in small buildings or rooms
Sample size	purchase costs	costs may preclude use of real-time monitors at many locations
Monitoring period	operating range	samplers used outdoors may not be applicable in both winter and summer
Contaminant sources	instrument range	numerous sources may require high range
Occupant density	obtrusiveness	large, noisy instruments may be unacceptable
Occupant age	device security	small children may tamper with devices
Occupant cooperation	mobility	occupants may be willing to wear personal monitors
Sampling locations	labor costs	maintenance, calibration, and transportation costs may be high if many remote sites are monitored

for sampling HCHO and particulate matter, and for maintenance and calibration of instruments. The laboratory has been used to monitor office buildings and residences [16,17].

Instrumentation in mobile laboratories can provide a wealth of real-time data to define spatial and temporal variations in contaminant concentrations if appropriate sample lines, signal cables, and monitors are configured into an automated system. Mobile laboratories can also be equipped for automated calibrations, thus allowing for unattended operation to a limited extent. However, the costs of obtaining this data can be high. The initial cost to equip a mobile laboratory is high because of the large number of sophisticated instruments and requirements for ancillary maintenance and calibration equipment. Operational costs are also high; a full-time technician is generally necessary. Labor costs can be substantial, particularly if the laboratory is moved frequently. Mobile laboratories are most suitable for programs requiring extended monitoring periods at a limited number of locations.

Real-Time Monitoring Instrumentation Placed within Buildings

An alternative approach to the mobile laboratory as a means to obtain real-time measurements indoors is to place the monitors in the building to be monitored. This approach is not without complications but is applicable in certain situations. If only a few buildings are to be monitored, this approach may be logistically feasible. Even if a larger number of buildings are to be monitored, this approach may be practical if the monitors are portable or only a limited number of contaminants or other parameters are monitored.

In a recent field survey, GEOMET configured a portable instrumentation package that could be easily transported for week-long samples in homes with unvented combustion appliances. The package included a small, nondispersive infrared CO_2 monitor; a portable NO_2 monitor; portable CO monitors; power supplies for temperature/relative humidity probes; and a data logger. The cabinet into which the monitors were installed could be handled easily by two technicians. With this package, time-dependent data could be collected at an indoor site and, by use of multiple monitors for some contaminants, at an outdoor site.

When placing instrumentation in the environment to be monitored, the effect of the instrumentation on the environment and its occupants must be considered. Heat dissipated from the instruments may affect airflow patterns and mixing. Similarly, some types of instrumentation, for example monitors requiring compressed gases, are precluded from use indoors due to safety considerations.

In addition to considering instrument impact on the environment to be monitored and its occupants, it is necessary to consider the impact of the environment and its occupants on the instruments. For example, variations in temperature and humidity may affect instrument performance.

Passive Devices

Passive methods for sample collection have found widespread use in IAQ monitoring programs [1–3]. They are particularly attractive for large surveys because of their ease of use, small size, and generally low cost. Passive devices for NO_2 [18], HCHO [19], and Rn [20], and a number of other contaminants are currently available. For some pollutants, such as VOCs, passive collection devices are currently being developed and tested [21]. Availability of passive samplers has been summarized by Nagda, Rector, and Koontz [8]. Passive devices generally are not as precise or accurate as real-time monitors, but their performance is acceptable in many cases. Major considerations in the use of passive collection devices are (1) the minimum detection limit of the analysis system because this dictates the minimum length of the sampling period, (2) minimum face velocity, (3) temperature and water vapor effects, (4) selectivity and interferences, and (5) sampler placement. As new passive methods are developed and refined, it is anticipated that these devices will find even greater use in IAQ monitoring.

Staged or Tiered Approaches

Staged and tiered approaches to IAQ monitoring have been used in a number of past IAQ studies. These approaches frequently combine the use of passive samplers, active (integrated) samplers, and real-time monitors. In a tiered approach, passive or active (integrated) methods are often used for sample collection at all locations and more sophisticated methods, such as real-time monitors, are used concurrently at a subset of locations. With a staged approach, all locations are initially monitored with passive or active (integrated) samplers; at a later date more detailed measurements are made with real-time monitors. Two examples of these approaches are presented for illustrative purposes.

In a survey of 150 homes, GEOMET used a three-tiered monitoring approach combining passive and integrated methods with real-time monitors to measure NO_2, CO, CO_2, temperature, relative humidity, air exchange rates, and appliance use. The approach used for this study is summarized in Table 4. As indicated in the table, measurements in the largest

TABLE 4—*Three-tiered approach for monitoring homes with combustion appliances.*

Number of Homes	Contaminant/Parameter	Monitor/Method
	TIER 1	
157	NO_2	Palmes tubes
	CO	integrated sampler
	temperature	maximum/minimum thermometer
	air exchange	blower door
	building/occupant factors	survey questionnaire
	appliance use	activity log
	TIER 2	
32	NO_2	Palmes tubes
	CO	portable monitor and integrated sampler
	temperature/relative humidity	hygrothermograph and maximum/ minimum thermometer
	air exchange	perfluorocarbon tracers (PFTs) SF_6 dilution (integrated samples) blower door
	building/occupant factors	survey questionnaire
	applicance use	activity logs
	TIER 3	
16	NO_x	portable monitor
	NO_2	portable monitor and Palmes tubes
	CO	portable monitor and integrated sampler
	CO_2	stationary monitor
	temperature/relative humidity	portable monitor and maximum/ minimum thermometer
	air exchange	PFTs, SF_6, and blower door
	building/occupant factors	survey questionnaire
	appliance use	thermocouples/thermistors and activity logs

number of homes were made with passive devices. For the second tier, a subset of homes was monitored with passive devices and some real-time instrumentation. The smallest subset of homes was monitored with several real-time monitoring instruments in addition to passive devices. Additionally, in this subset, thermocouples and thermistors were used to monitor combustion appliance use. Survey questionnaires for recording various factors were used in all homes. As part of the quality control program for this study, passive devices were colocated with real-time monitors in Tier 3. As part of the QA program, results of measurements with thermocouples and thermistors were compared to occupant entries on appliance use logs as an indicator of the accuracy and completeness of these survey instruments.

Nitschke et al. [22] used a staged approach in a recently completed study. They conducted an initial survey with blower door tests for air leakage and radon measurements with passive samplers in 60 homes. In 30 homes with combustion appliances, passive or integrated methods were used to measure air exchange rates, NO_2, CO, HCHO, and respirable particles (RSP). Follow-up monitoring was conducted in four homes with real-time measurements of Rn, RnP, RSP, air exchange rates, temperature, humidity, and outdoor meteorology. In six houses, real-time measurements of combustion pollutants in the follow-up stage included monitors for CO, CO_2, NO_x, NO_2, RSP, temperature, humidity, air exchange rates, and outdoor meteorology. Logs for describing indoor pollution activity were also included.

Staged or tiered approaches are an attractive alternative to using mobile laboratories because cost per monitoring location is much lower. Compared to the mobile laboratory, this approach can be used for monitoring a larger number of locations. Detailed temporal data can be obtained for dwellings of the greatest interest. Because passive samplers can be colocated with more sophisticated real-time monitors, these approaches also provide a means for evaluating the performance of passive devices for QA purposes under realistic sampling conditions.

Concluding Remarks

Researchers designing IAQ monitoring programs have a wide range of measurement methods and monitors available for use. The potential applicability of available measurement devices for use indoors must, however, be carefully evaluated based on the design of the monitoring program and the operating and performance characteristics of the measurement devices.

Although considerable progress has been made in recent years in the adaptation of existing measurement methods and development of new methods, further research and development are needed in this area. A major effort needs to be undertaken to standardize IAQ measurement methods. Standardized IAQ monitoring protocols, sample collection and analysis methods, and methods for data summary and interpretation can greatly improve the value of results.

References

[1] *Indoor Pollutants,* National Research Council, Committee on Indoor Pollutants, National Academy Press, Washington, DC, 1981.
[2] *Environment International,* Vol. 8, Nos, 1–6, 1982, pp. 1–534.
[3] *Indoor Air,* B. Bergland, T. Lendvall, and J. Sundall, Eds., *Proceedings,* 3rd International Conference on Indoor Air Quality and Climate, Swedish Council for Building Research, Stockholm, 1984.
[4] Wallace, L. A. and Ott, W. R., *Journal,* Air Pollution Control Association, Vol. 32, No. 6, June 1982, pp. 601–610.
[5] *Indoor Air Quality Handbook,* Sandia Report SAND82-1773, Sandia National Laboratories, Albuquerque, NM, 1982.

[6] Hawthorne, A. R., Matthews, T. G., and Vo-Dinh, T. in *Indoor Air Quality*, P. J. Walsh, C. S. Dudney, and E. D. Copenhaver, Eds., CRC Press, Inc., Boca Raton, FL, 1984, pp. 15–53.

[7] Nagda, N. L. and Rector, H. E., "Guidelines for Monitoring Indoor Air Quality," NTIS PB83-264465, National Technical Information Service, Springfield, VA, 1983.

[8] Nagda, N. L., Rector, H. E., and Koontz, M. D., *Guidelines for Monitoring Indoor Air Quality*, Hemisphere Publishing Corporation, New York, 1986.

[9] Harrje, D. T., Grot, R. A., and Grimsrud, D. T. in *Building Design for Minimum Air Infiltration, Proceedings*, 2nd AIC Conference, Air Infiltration Center, Berkshire, UK, 1980, pp. 113–133.

[10] *1985 Annual Book of Standards, Volume 11.03, Atmospheric Analysis; Occupational Health and Safety*, American Society for Testing and Materials, Philadelphia, 1985.

[11] *Air Sampling Instruments for the Evaluation of Atmospheric Contaminants*, 6th Ed., American Conference of Governmental Industrial Hygienists, Cincinnati, 1983.

[12] Linch, A. L., *Evaluation of Ambient Air Quality by Personnel Monitoring*, Vols. I and II, 2nd ed., CRC Press, Boca Raton, FL, 1981.

[13] Koontz, M. D. and Nagda, N. L., "Systematic Development of Survey Instruments for Indoor Air Quality Studies," presentation at the 78th Annual Meeting of the Air Pollution Control Association, Detroit, 16–21 June, 1985.

[14] "Interim Guidelines and Specifications for Preparing Quality Assurance Project Plans," U. S. EPA Report No. EPA-60014-83-004, NTIS PB83-170514, U. S. Environmental Protection Agency, Springfield, VA, 1983.

[15] Grimsrud, D. T., Hodgson, A. T., Turk, B. H., Koonce, J. F., and Girman, J. R. in *Indoor Air*, B. Berglund, T. Lindvall, and J. Sundell, Eds., Swedish Council for Building Research, Stockholm, Vol. 4, 1984, pp. 215–220.

[16] Nagda, N. L., Koontz, M. D., and Rector, H. E., "Energy Use, Infiltration, and Indoor Air Quality in Tight, Well-Insulated Residences," EPRI EA/EM-4117, Research Project 2034-1 Final Report, Electric Power Research Institute, Palo Alto, CA 1985.

[17] Moschandreas, D. J., Zabransky, J., and Pelton, D. J., "Comparison of Indoor and Outdoor Air Quality," EPRI EA-1733, Project 1309 Final Report, Electric Power Research Institute, Palo Alto, CA, 1981.

[18] Palmes, E. D., Gunnison, A. F., DiMattio, J., and Tomczyk, C., *Journal*, American Industrial Hygiene Association, Vol. 37, Oct. 1976, pp. 570–577.

[19] Geisling, K. S., et al., *Environment International*, Vol. 8, Nos. 1–6, 1982, pp. 153–158.

[20] George, A. C., *Health Physics*, Vol. 47, 1984, pp. 867–872.

[21] Coutant, R. W., Lewis, R. G., and Mulik, J. D., *Analytical Chemistry*, Vol. 58, 1986, pp. 445–448.

[22] Nitschke, I. A., Traynor, G. W., Wadach, J. B., Clarkin, M. E., and Clarke, W. A., "Indoor Air Quality, Infiltration, and Ventilation in Residential Buildings," NYSERDA Report 85-10, New York State Energy Research and Development Authority, Albany, 1985.

Elia M. Sterling,[1] *Edward D. McIntyre,*[1] *Christopher W. Collett,* [1]
Jack Meredith,[2] *and Theodor D. Sterling*[3]

Field Measurements for Air Quality in Office Buildings: A Three-Phased Approach to Diagnosing Building Performance Problems

REFERENCE: Sterling, E. M., McIntyre, E. D., Collett, C. W., Meredith, J., and Sterling, T. D., **"Field Measurements for Air Quality in Office Buildings: A Three-Phased Approach to Diagnosing Building Performance Problems,"** *Sampling and Calibration for Atmospheric Measurements, ASTM STP 957,* J. K. Taylor, Ed., American Society for Testing and Materials, Philadelphia, 1987, pp. 46–65.

ABSTRACT: This paper reports on a phased system of integrated measurements of ventilation and occupant complaints that is relatively inexpensive yet may lead to effective diagnosis of a building's indoor air quality (IAQ) problems.

1. Physical inspection of the building, surroundings, and ventilation system.
2. Use of standardized questionnaire to (1) compare distribution of responses to those obtained from buildings without increased occupant complaints, (2) determine major types of complaints, (3) inspect distribution of complaints by location in building, and (4) check for specific complaints that may indicate special sources of contaminants.
3. Use carbon dioxide (CO_2) concentrations as an index of overall ventilation adequacy.
4. Use carbon monoxide (CO) distribution as an index for overall infiltration pattern of combustion byproducts.
5. Measure fresh air supply by location, during variation of mechanical system operation, using tracer gases, smoke pencils, and tests of actual functioning of the mechanical system.
6. Compare design specification of mechanical system with all observations.

Specific measurement techniques are illustrated through a case study.

KEY WORDS: air quality, calibration, sampling, atmospheric measurements, testing, ventilation, office buildings, measurement, human factors, heating, health, environment, contamination, comfort, air conditioning, indoor air quality, building performance

The term "Sick Building" is often used to describe buildings in which a large number of occupants experience symptoms of "Tight Building Syndrome" for which no specific cause can be determined. In general, investigators have concentrated on extensive monitoring of various indoor pollutants and environmental parameters, while, for the most part, building systems have been assumed to be functioning as designed. Typical of investigations of this

[1] Director of Building Research and senior research associates, respectively, Theodor D. Sterling Limited, 70–1507 West 12th Avenue, Vancouver, B. C., Canada V6J 2E2.
[2] Manager, Energy Conservation, British Columbia Buildings Corporation, Victoria, B. C. Canada V8W 2T4.
[3] Professor, School of Computing Science, Simon Fraser University, Burnaby, B. C., Canada V5A 1S6.

type is a large Federal government office complex in Hull, Quebec, that houses 5300 public servants of which 49% experience nose irritation, 46% eye irritation, and 83% headaches. After taking extensive measures of indoor air quality over the course of one year, researchers still could not identify any specific causal agent nor could they recommend effective control measures [1].

Many sick buildings have now been studied by qualified investigators. Although most studies were inconclusive, there now exists a substantial archive of data in the form of both published and unpublished reports. These data include such parameters as air quality, ventilation, lighting, acoustics, and reported effects on the health and comfort of occupants as well as research protocol and instrumentation. Much can be learned from a review of these reports regarding what has been done and found by the investigators. Such a review was recently completed. Nearly all written reports of these sick building investigations (nearly 500) were collected, and data from the reports were loaded into a computerized Building Performance Database (BPD), [2,3].[4] Based on review of these data, in addition to experience gained from numerous investigations undertaken in Canada, the United States, and Great Britain, a practical and systematic strategy has now been developed and tested on how to proceed or where to look to diagnose a sick building, identify the cause of problems, and prescribe a course of action designed to correct the situation [4].

The systematic strategy is essentially a screening process that can be separated into three more-or-less distinct phases.

Phase 1

(*a*) Administration of a Work Environment Survey Questionnaire to all building occupants as a means of documenting environmental problems and symptoms reported in the building and to locate clusters of complaint areas for detailed monitoring.

(*b*) General overview of the performance of the existing mechanical systems including site inspections and review of plans.

Phase 2

(*a*) Measurement of carbon dioxide as an indicator of a build-up of contaminants generated indoors from occupants and equipment.

(*b*) Measurement of carbon monoxide as an indicator of combustion byproducts infiltrating from the outside, especially from parking garages or from other sources of indoor combustion.

Phase 3

(*a*) Measurement of air exchange rates with tracer gas to determine the total ventilation rate within the building, as well as the pattern of air movement through the building including air leakage from potential indoor pollutant sources such as parking garages.

Such a phased approach may often eliminate the need for extensive measurements by either quickly identifying obvious problems (such as ventilation system malfunctions) or focusing the investigation attention on the most likely causes of health and comfort problems.

---

[4] Information about use of and access to the Building Performance Database can be obtained by contacting the author at Theodor D. Sterling Ltd.

For demonstration purposes, one complete building evaluation is used here to illustrate the methods used.

The Buildings

The study building is a 10-year-old, four-story, sealed, mechanically ventilated and air conditioned office building in Victoria, British Columbia. The building is located over a partly underground parking garage. Tempered, ventilating air is supplied mechanically by means of a central air conditioning system for the core office areas, and by individual heat pump units located under every window for the perimeter office areas. The heat pump units are directly ducted to the outdoors with the exception of six units located on the first floor that draw outdoor air in through a common duct running through the basement parking garage.

A control building was used in this evaluation. The control building is also a four-floor office building. However, it is approximately 50 years old and is equipped with operable windows for ventilation and radiators for heat. Each building is occupied by approximately 130 government office employees. Although for purposes of investigation, a sealed, mechanically ventilated control building more like the study building may have been more appropriate, this building was selected as a control because:

(a) it is the same size and height,
(b) it houses the same government department,
(c) it is directly across the street, and
(d) it does not have a history of air quality complaints.

Phase 1: The Survey of Occupants' Perceptions

(A) *Administration of Questionnaire*

A standardized survey questionnaire was administered to all occupants of both the study and control building (see Appendix A). The questionnaire contains a large number of questions on environmental, health and comfort, stress, and other conditions and has been administered to nearly 4000 office workers across North America and England. Baseline data has been indexed and calibrated [5,6]. The response rate from both buildings was over 90%. Survey results identified a predominance of complaints of inadequate ventilation and dry and stuffy air among building occupants of the study building compared to the control building. The symptoms reported most prevalent among occupants of the study building were of mucous membrane irritation such as the eyes, nose, and throat. Symptoms also included headaches and fatigue. On the other hand, there were no increases in symptoms less likely affected by air quality such as fatigue due to posture and seating problems, indicating that respondents tended to supply specific information rather than express general complaints. Of the four floors of the study building, the third floor was found to have the highest incidence of both environmental complaints and symptoms.[5]

(B) *Overview of Mechanical Systems*

The study building has two systems for the introduction of outdoor air to the office floors. The primary air supply to the core area offices is a constant volume system that provides

---

[5] Because of the large number of items pertaining to the questionnaire, the results of these analyses are not further summarized here. However, a copy of the complete report may be obtained from the British Columbia Buildings Corporation (see Ref 7).

approximately 2000 L/s of outdoor air that is distributed in equal quantities between the four office floors. The outdoor air duct delivers fresh air directly from the rooftop mechanical penthouse to the center of the ceiling plenum space on each floor. Ceiling-mounted heat pumps then draw a mixture of outdoor and return air from the ceiling plenum space, temper it, and diffuse the mixed air into the office area.

A second system provides a mixture of outdoor and return air to the perimeter office space through 88 perimeter console fan coil units. The perimeter console units are located below the windows on three sides of the building. All console units draw air directly from the outdoors, except six located along the east-facing side of the first floor. These six units draw outdoor air from a duct that passes through the basement parking garage.

There are two building exhaust systems. Washrooms and photocopier rooms are exhausted through the roof. The capacity of this exhaust is approximately 1000 L/s. Before leaving the mechanical penthouse, this exhaust air passes through an air-to-air heat exchanger that transfers heat from the exhaust to the incoming fresh air.

A second exhaust system draws air from the occupied office space and delivers it through a "heating chase" into the first aid room and stair lobby of the basement parking garage. This system exhausts approximately 1800 L/s from the general office areas and supplies it to the basement parking garage. There is also a direct supply of approximately 100 L/s of outdoor air to the parking garage and an undetermined supply through the open garage door.

Both the basement parking garage and rooftop elevator penthouse have separate exhausts.

Phase 2: Environmental Measurements

Carbon monoxide (CO) and carbon dioxide (CO_2) were measured two times a day, once in the morning and once in the afternoon, on work days during a continuous three-week period. Measurements were taken in the study building at a number of sites on each floor selected as typical of both open plan and enclosed office space (eight on Floor 1, seven on Floor 2, six on Floor 3, seven on Floor 4, and one each on the roof and in the garage). Measurements were taken at selected locations in the office space on two floors of the control building. All outdoor sampling was done on the roof of the study building. Because of the close proximity of the two buildings (directly across a narrow street), ambient outdoor conditions were assumed to be similar for the control building.

Carbon monoxide and carbon dioxide were measured using a Wilks Miran 1A ambient air Analyzer with a reported error of ± 2%. The equipment was calibrated using known concentrations of both gases. Measurements for both CO and CO_2 were made in 32 office locations in the study and two in the control building at approximately seated breathing-zone height. They were also taken on the roof, and CO was measured in the basement parking garage of the study building. Measurements were used to determine:

1. the effectiveness of the ventilation system in removal of indoor-generated contaminants, and
2. the penetration of outdoor and parking garage-generated contaminants.

Carbon Monoxide

Table 1 shows mean CO concentrations measured before and after lunchtime in the occupied office space of the study building as well as the basement parking garage, outdoors on the roof, and in the occupied office space of the control building (CO values were almost identical for all sites per floor so that only one measurement is shown of mean concentrations

TABLE 1—*Mean carbon monoxide concentrations and numbers of observations in parts per million by floor for study building, including basement, control building, and outdoors.*

	Study Building						Control Building	
	Basement Parking Garage	Floor 1	Floor 2	Floor 3	Floor 4	Roof	Floor 3	Floor 4
a.m.	11.4	8.9	7.8	8.2	7.8	8.7	6.8	7.5
	(14)[a]	(30)	(15)	(15)	(15)	(15)	(15)	(15)
p.m.	15.3	7.9	7.5	7.8	7.4	5.9	7.3	7.6
	(13)	(30)	(15)	(15)	(15)	(15)	(15)	(15)

[a]Numbers of observations.

for each floor in Table 1). The exact times of measurement were not used for evaluation purposes, however, the mean concentrations of CO do not vary greatly by floor or by morning or afternoon. The highest CO concentrations were found in the basement parking garage, with morning and afternoon means of 11.4 and 15.3 ppm, respectively. The CO concentration on all floors of the study building were similar and did not vary significantly by time of day. Although not statistically significant, CO concentrations were slightly lower in the control building office areas.

Daily variations of CO concentrations measured in the study building, are shown in Fig. 1. Both mean morning and mean afternoon levels shown in the graph indicate an association between indoor, outdoor, and basement concentrations. These associations are confirmed by an analysis using Pearson Product Moment Correlation Coefficients. The CO concentrations outdoors and in the garage where each matched to the indoor CO value taken most closely in time but following measures of CO taken on the roof and in the garage. Correlations were then computed for a.m. and p.m. separately. Correlations are given in Table 1. The relationship between indoor and outdoor, and indoor and basement CO are both statistically significant at the 5% level.

For indoor/outdoor CO

$$\text{a.m. } r = 0.381 \qquad p \leq 0.015 \qquad \text{degrees of freedom (df)} = 14$$

$$\text{p.m. } r = 0.389 \qquad p \leq 0.02 \qquad df = 14$$

$$\text{a.m. } r = 0.284 \qquad p \leq 0.04 \qquad df = 13$$

$$\text{p.m. } r = 0.317 \qquad p \leq 0.02 \qquad df = 12$$

An analysis of variance also confirms differences between floors (although with only marginal statistical significance ranging from $p \leq 0.2$ to $p \leq 0.1$).

Figure 2, shows the distribution of CO in 209 similar study areas (updated from Ref 8). In absolute terms, the CO concentrations encountered in both buildings are high in comparison with those measured in similar buildings by the U. S. Public Health Service, the Department of Energy, the Environmental Protection Agency, and various universities and research institutes. Outdoor levels measured in these studies are not given in Fig. 2, however, they were reported before as similar to indoor levels [9]. The buildings found in both the study and control buildings fit into the high end of the measured range. It should be kept

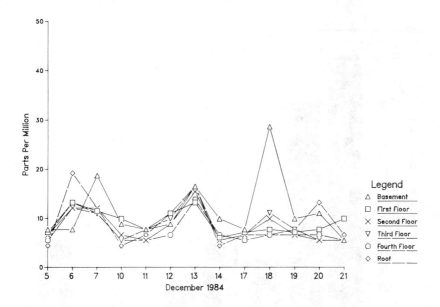

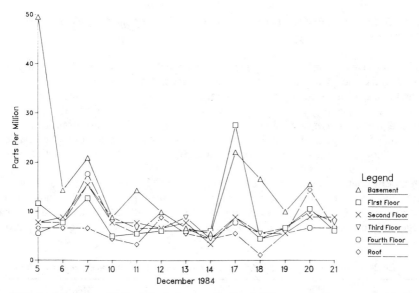

FIG. 1—*Mean morning* (top) *and afternoon* (bottom) *CO concentrations measured in the study building.*

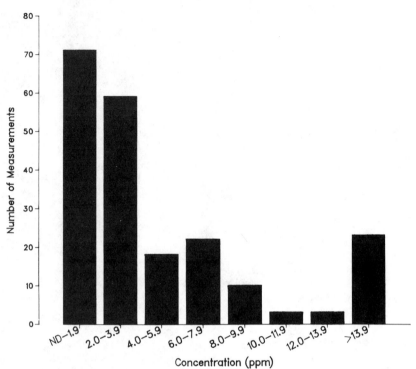

FIG. 2—*Distribution of 209 CO measurements extracted from the building performance database.*

in mind, however, that CO values in Fig. 2 were obtained in various locations using different measuring techniques and are thus only approximately comparable.

The finding from the survey questionnaire, administered in Phase I, showing substantially fewer complaints among occupants of the control building, seems to rule out CO as the only source of problems in the study building. The relatively high CO concentrations encountered in the study building could either have been the result of penetration of automobile exhaust from outdoors or the basement parking garage. Because smoking is prohibited throughout the study building, this source was ruled out.

In order to determine the relative effect on indoor CO levels of both garage and outdoor sources, two sets of correlations were computed. Table 2 gives the correlation coefficients between indoor CO separately with outdoor CO and with CO in the basement parking garage for each floor, computed as previously described. The correlation between outdoor

TABLE 2—*Correlations on indoor CO with outdoor CO and basement (garage CO) by floors.*

	Floor 1	Floor 2	Floor 3	Floor 4
Correlations of indoor CO				
With outdoor CO[a]	0.48	0.56	0.60	0.50
With basement CO[b]	0.40	0.33	0.28	0.17

[a]For all rs, df = 29.
[b]For all rs, df = 26.

TABLE 3—*Multiple correlation and normalized regression coefficients of CO contribution from outdoors and basement (garage) for each floor.*

	Floor 1	Floor 2	Floor 3	Floor 4
Multiple r^a	0.67	0.69	0.70	0.55
Regression coefficients				
Roof	0.53	0.61	0.64	0.53
Basement	0.46	0.40	0.35	0.23^b

[a]All coefficients are statistically significant with $p \leq 0.05$, except this coefficient.
[b]For all multiple rs, df = 25.

CO and indoor CO is approximately the same from one floor to the next but correlations between CO in the garage and on the floors decrease progressively. All the correlations between CO indoor and outdoor are significant for the first and second floors ($p \leq 0.01$), and those with the basement are also significant ($p \leq 0.05$). But, more important than statistical significance is the trend showing the basement parking garage as a significant source of indoor CO.

In order to further assess the contribution of the basement to whole building CO levels, Table 3 presents computed multiple correlation coefficients combining the effects of garage and outdoor CO levels. The first line of Table 3 gives the multiple correlation coefficients. Of more interest than these correlations are the normalized regression coefficients (also known as weights or beta coefficients) for roof and basement, given in the following two lines. These beta coefficients demonstrate the relative size difference between the regression coefficient or the beta value for the roof and for the basement. Again, note that while the beta coefficient for the roof remains relatively similar, the basement values decline with distance from the basement. The basement parking garage appears to make a significant contribution when evaluated independently up to the third floor. The fourth floor contribution falls short of statistical significance.

These results verify that there is a penetration of CO and other exhaust products from the basement parking garage that is sufficiently significant to be picked up by statistical tests up to the third floor. If CO penetrates the building, so do other fumes from automobile traffic in the parking garage, such as volatized gasoline vapors. The garage thus emerges as a major source of air contaminants.

Carbon Dioxide

Carbon dioxide measurements in the occupied space of the study and control buildings and on the roof of the study building are shown in Table 4 (because CO_2 varied by site

TABLE 4—*Mean CO_2 concentrations and number of observations in parts per million by floor for study building, control building, and outdoors.*

	Study Building					Control Building	
	Floor 1	Floor 2	Floor 3	Floor 4	Roof	Floor 3	Floor 4
a.m.	434.9	468.4	516.2	509.6	255.6	533.3	483.3
	$(106)^a$	(103)	(84)	(104)	(15)	(15)	(15)
p.m.	471.2	491.6	545.5	509.6	288.9	408.9	433.3
	(104)	(105)	(86)	(102)	(15)	(15)	(15)

[a]Number of observations.

sufficiently, means in Table 4 are based on all observations on each floor). Comparing CO_2 measurements taken in the morning and afternoon with CO measurements taken in the morning and afternoon (Table 1) at the same locations in the occupied office space shows that there is a larger floor-by-floor and diurnal variation for CO_2 than for CO. Two trends are apparent in the study building;

1. CO_2 tends to be higher on the upper two floors, and
2. CO_2 concentrations increase in the afternoon in comparison to the control building where concentrations actually decreased on both the third and fourth floors in the afternoon.

The patterns of daily variation and variation between floors for CO_2 are shown in the graphs of Fig. 3. There are sharp contrasts between the pattern of morning and afternoon CO_2 concentrations during the study period as well as differences in CO_2 between floors. The graph of morning concentrations of CO_2 shows a strong positive relationship between indoor and outdoor levels. However, this relationship is not evident for afternoon CO_2 concentrations.

The observed differences in the indoor/outdoor relationship between morning and afternoon concentrations of CO_2 are verified by the large difference in values of the correlation coefficients (computed as those for CO).

$$\text{a.m. } r = 0.484 \qquad p \leq 0.0001 \qquad df = 14$$

$$\text{p.m. } r = 0.028 \qquad p \leq 0.79 \qquad df = 14$$

In the mornings, indoor CO_2 concentrations are strongly influenced by the outdoor levels as can be seen from the large correlation between indoor and outdoor concentrations. However, this changes over the day, with outdoor CO_2 no longer a main determinant of indoor CO_2 in the afternoon (and has decreased to zero for all practical purposes).

The significance of variations in CO_2 by floor, time, and day were further verified by an analysis of variance. Differences between floors were found to be highly significant ($p \leq 0.001$) and so were differences for time of day ($p \leq 0.001$). The results emphasize the consistent increase of CO_2 in the afternoon with the highest levels on the third floor.

For all these reasons, diurnal variations in CO_2 may not have a great impact. First, CO_2 levels decrease in the control building; second, while there is a correlation between outdoor and indoor CO_2 in the morning, this correlation disappears in the afternoon; and, third, there is a floor effect as well.

Figure 4, extracted from the Building Performance Database, shows the distribution of CO_2 values measured by the U. S. Public Health Service, the Department of Energy, the Environmental Protection Agency, and various universities and research institutes in 100 similar study areas (updated from Ref 8). The CO_2 concentration measured in the study building is similar to that measured in other buildings.

Phase 2 Screening Process Discussion

In comparison with other study buildings, both the study and control building have higher than average CO concentrations. Concentrations of CO in the office space of the study building are affected by both outdoor sources and fumes from the parking garage in the basement. The outdoor contribution to indoor CO is consistent throughout the building. The CO from the parking garage, although permeating the whole building, is strongest on

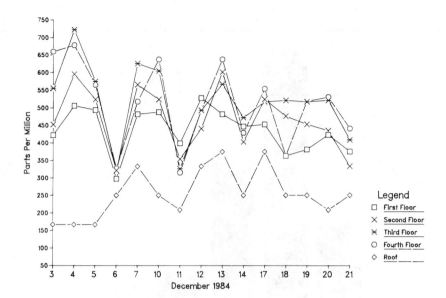

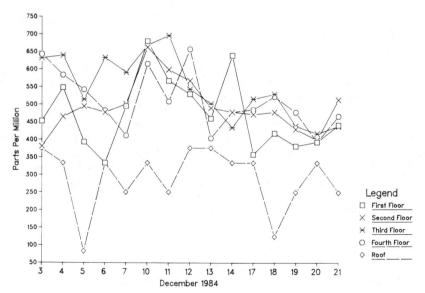

FIG. 3—*Mean morning* (top) *and afternoon* (bottom) CO_2 *concentrations measured in the study building.*

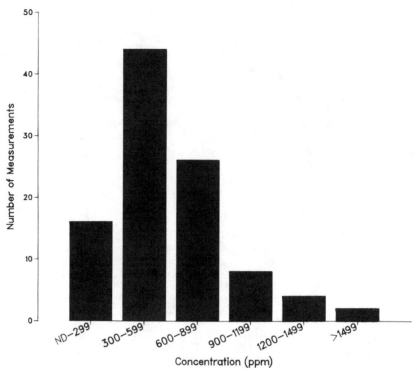

FIG. 4—*Distribution of 100 CO_2 measurements extracted from the building performance database.*

the first floor and progressively weakens by floor level. Insofar as CO is an indicator of vehicle exhaust fumes, the conclusion is compelling that part of the indoor problem is created by the many irritants in exhaust gases and aerosols from the garage.

Carbon dioxide concentrations show higher variability by floor, day, and time than carbon monoxide. Results show that outdoor CO_2 has a strong effect on indoor concentrations in the morning, but that effect is no longer present in the afternoon after the building has been occupied. Even though the mean CO_2 concentration measured in the afternoon is below the 600 ppm level considered acceptable [10], one interpretation fitting these observations, is that the ventilation system in the study building is unable to keep pace with occupant generated CO_2 and, by extension, that the ventilation system is unable to keep pace with accumulation of other contaminants in the building, including those from the garage.

Phase 3: Ventilation System Performance

Phase 3 consisted of a number of tests and evaluations performed to:

1. measure the amount of air being provided to the building occupants by tracer gas and smoke pencil evaluations, and
2. determine the path by which polluted air from the basement parking garage enters the occupied office floors, again by tracer gas and smoke pencil evaluations.

Measurement and Evaluation of Outdoor Air Supply

Tracer gas was used to measure the actual outdoor air supplied to the entire building. Sulfur hexafluoride (SF6) was introduced into the study building through the rooftop outdoor air intake upstream of the air-to-air heat exchanger. The quantity of SF6 released was not determined. The concentration of SF6 in the rooftop exhaust air was then measured using a Miran 1A infrared gas analyzer and strip chart recorder set up in the mechanical system penthouse. The collection tube attached to the Miran 1A was inserted into the building's main exhaust duct. The decay of concentrations of SF6 in the exhaust air was continuously recorded. Although the pathway to occupants is not shown by this method, it provides a measure of the outdoor air provided to building occupants under two modes of ventilation system operation:

1. with the perimeter consoles (fan coil units) operating under normal conditions (as found at the end of the workday when the investigators arrived), and
2. with all perimeter consoles turned off.

The tracer gas decay method was also used to determine the actual amount of outdoor air provided to occupants of each floor and to the first floor conference room.

In addition, the percentage of outdoor air provided by two of the 88 perimeter console units was determined by monitoring tracer gas concentrations in the ambient room air and in the air supplied to the room by the perimeter units (directly at the supply grill again using the Miram 1A gas analyzer). The difference between the concentration of tracer gas in the ambient room air and the air supplied by the perimeter units provides a measure of the percentage of outdoor air delivered by each unit.

Movement of air currents in the building were visually determined by following the movement of "smoke" produced by smoke pencils.

Tracer gas tests conducted of ventilation system performance, flow measurements of the system that had previously been undertaken, and the design flows specified on the mechanical drawings are presented in Table 5. The outdoor air flow per person was calculated using both the American Society of Heating, Refrigerating and Air Conditioning Engineers (ASHRAE) design estimate of seven persons/100 m^2 of office floor space, and the actual or official population count.

Tracer gas testing shows that perimeter console units, when operated under normal conditions, provide approximately 30% of the outdoor air supplied to building occupants. With

TABLE 5—*Tests of outdoor air supplied to the study building.*[a]

Tracer Gas Tests	Total Outdoor Air Flow	Outdoor Air per Person ASHRAE[b]	Outdoor Air per Person Actual[c]
Normal operation	3121	14	21
Perimeter consoles off	2057	9	14
Flow measurement tests	2085	9	14
Mechanical design specifications	1915	9	13

[a]In litres per second, L/s.
[b]Based on ASHRAE estimates of maximum population density seven persons/100 m^2. This gives an estimated population of 224 (ASHRAE, 1981).
[c]Based on population of 149.

TABLE 6—*Tracer gas tests of outdoor air supplied to each floor of the study building.*[a]

	Total Outdoor Air Flow	Actual Population	Outdoor Air Supply per Person
First floor	630	45	14
Second floor	737	33	22
Third floor	800	43	18
Fourth floor	676	28	24

[a]In litres per second, L/s.

the console units turned off, the supply of outdoor air to occupants on a per person basis (using both ASHRAE and actual population estimates) is 14 L/s person (L/s/p) compared to the ASHRAE Standard of 10 L/s/p (Ventilation for Acceptable Air Quality, ANSI/ASHRAE Standard 62–1981). However, with the consoles in normal operation the outdoor air supply is well in excess of the ASHRAE standard.

The close agreement between the tracer gas tests, the flow measurements, and the design specifications in terms of outdoor air supply shows consistency between the different methods of determining ventilation system performance (that is, tracer gas and flow measurements). Further, the close agreement between design and measured values indicate that the ventilation system is performing as intended.

Table 6 shows the air supply on each floor (based on areas calculated from plans) and the air supply per person in L/s (based on the actual population of each floor determined as in Table 5). Ventilation rates for each floor comply with the current ASHRAE standard of 10 L/s/p.

Table 7 presents results of testing the amount of outdoor air provided by two of the 88 perimeter fan coil console units operated at the maximum setting. The total air supplied by each console unit is approximately 100 L/s on maximum setting and the fresh air component is 18 to 20 L/s. Then assuming that all 88 console units supply a similar quantity of outside air and if all were operating at maximum capacity, they would supply approximately 1760 L/s/p of outdoor air to the building. If necessary to improve the comfort of the building occupants, the outdoor air supplied to the building could therefore be substantially increased by operating all perimeter units at maximum setting.

Table 8 shows that while the supply of outdoor air to the first floor as a whole is adequate, the supply of outdoor air to the conference room on the first floor (presuming a population of 20 people) is inadequate. The ventilation system providing air to the conference room is a constant volume system that does not vary the amount of air delivered according to occupancy. At times of low occupancy, the system provides adequate ventilation, however, the room is intended for large meetings of 20 or more. While the ASHRAE standard suggests a supply of 17.5 L/s/p of outdoor air, the existing system supplies only 3.5 L/s/p (based on an occupancy of 20 people indicated by seating provided). The inadequate air supply found for this conference room, indicated that the mechanical design documentation warranted closer inspection. After careful review of the plans, other densely populated areas of the building were also found to be poorly supplied with fresh air.

TABLE 7—*Tracer gas tests of outdoor air supply from two console units.*

	Unit 1	Unit 2
% Outdoor air supplied by console units	18%	22%

TABLE 8—*Outdoor air supply.*[a]

	Total Outdoor Air Flow	Outdoor Air Supply/Person	Air Change/h
Conference room	69	3.5[b]	1.5
First floor	630	14.0	0.6

[a]In litres per second, L/s.
[b]Based on 20 persons (ASHRAE Standard for a conference room is 17.5 L/s/p).

Qualitative tests of air circulating and mixing patterns made visually using smoke pencils showed good mixing of air in the occupied office space.

Determination of Pollutant Pathways from the Garage

Carbon monoxide measurements collected as part of Phase 2 of the three-phase screening process showed a strong correlation between the concentrations of CO measured in the garage and on the first floor. A sheet metal duct providing outdoor air to the six perimeter console units located along the east side of the first floor was suspected to be the pathway for the passage of CO and other pollutants from the garage into the building. This duct passes through the garage and was found by smoke pencil tests to entrain air from the garage when the consoles were operating.

To test whether this duct was providing air from the parking garage to the occupied office space through the console units, tracer gas, nitrous oxide (N_2O), was released in the garage adjacent to the duct. The concentration of tracer gas penetrating the first and second floor was then measured using a Miran 1A gas analyzer. Table 9 shows that tracer gas was detected throughout the first floor, with higher concentrations on the east side near the six consoles supplied with outdoor air by the duct passing through the garage. No tracer gas, however, had penetrated into the second floor. These results show that the duct supplying the console units is the primary path by which contaminated air from the garage enters the first floor office space.

Discussion and Conclusion

The study of air quality parameters in office buildings not only can be a costly undertaking, but because of the nature of the work setting and the different interests involved (that is, owner, tenant, worker), access for monitoring is often difficult to arrange. For these reasons,

TABLE 9—*Tracer gas testing of leakage patterns from the parking garage.*

	Nitrous Oxide Levels	
	First Readings[a]	Second Readings[a]
Parking garage	3.52	2.40
First floor east	0.68	0.65
First floor west	0.32	...[b]
Second floor	...[b]	ND[c]

[a]Readings are in absorption units and are only relative.
[b]No data collected.
[c]Nondetectable.

there are relatively few office buildings in which a comprehensive evaluation of air quality parameters has been completed.

The studies that have been completed have often had conflicting foci or objectives; namely,

1. to explore the complexity of factors that determine the quality of a building's performance, and
2. to locate the major problems in a particular building with poor air quality in the most economical way possible so that satisfactory working conditions may be created.

Clearly, the procedures adopted and tested here address themselves to the second objective. The purpose behind our phased approach is to converge on the factors that make for poor air quality in a particular problem building. The plans of the building, as thoroughly studied, and the perceptions of building occupants are used as a guide for what for and where to search. Changes in CO_2 by time and location indicate the adequacy of the mechanical system, while CO levels in different locations indicate possible pathways and levels of infiltrated contaminants from outdoors. Where to sample CO_2 and CO as well as sampling schedules are determined by a combination of familiarity with building plans and occupant perceptions. Tracer gas and visualizations of airflow give final information of pathway and spread of contaminants from source to specific target areas. They also give a final check on the extent to which the existing air control system fulfills the design aims of the specified system.

The phased approach described here is most useful if a building's problem stems from a source of combustion (garage, photocopier, poorly located air intakes). But the phased approach is not exclusively aimed at combustion sources. For instance, a high incidence of complaints about "fever" and "headaches," coupled with a relatively low incidence of eye and respiratory irritation, and possibly also a low level of complaints about dryness, would direct attention toward biological factors such as bacteria or fungi in humidifiers or penetrating from outside. What we suggest is that any phased approach is not cast in stone but should be used as a point of departure. Also, a factor for consideration is the fact (established by extensive reviews of building illness problems) that the vast majority of these problems are caused by faulty ventilation often paired with infiltration from combustion byproducts [11,12]. The phased approach outlined here will be most effective in locating precisely these same causes of unsatisfactory indoor air quality.

APPENDIX

OFFICE WORK ENVIRONMENT SURVEY

© TDS LIMITED, 70-1507 West 12th Ave., Vancouver, British Columbia, Canada V6J 2E2

This survey is being used to determine the quality of your work place. Your assistance in completing the following questions as accurately as possible is very much appreciated.

All information will be treated as confidential and anonymous.

MARKING DIRECTIONS

- DO NOT staple or tape this sheet
- DO NOT make extraneous marks
- USE a soft lead pencil (e.g. HB)
- FILL IN the circle completely
- ERASE COMPLETELY any answer you wish to change

CORRECT

INCORRECT

FOR OFFICE USE ONLY

JOB	DEPT	BLDG	FLOOR

GROUND
BASEMENT

2. Marital Status
- ○ Single
- ○ Married
- ○ Separated
- ○ Divorced
- ○ Widowed

3. Sex
- ○ Male
- ○ Female

4. Number of children
- ○ None
- ○ One
- ○ Two or more

5. Highest level of Educational Attainment (if applicable)
- ○ CSE/O Level
- ○ A Level, or High School Graduation
- ○ Technical Training
- ○ College Education
- ○ Undergraduate Degree
- ○ Postgraduate Degree

ALL RESPONDENTS, PLEASE ANSWER THE FOLLOWING QUESTIONS.

PLEASE WRITE ONLY WITHIN THE SHADED BOX

JOB TYPE

DEPARTMENT

BUILDING FLOOR

6. Date of starting work in building

ERIBC Testing Service - Forms Development & Design 84-04

1. Date of Birth

MONTH	DAY	YEAR
JAN		
FEB		
MAR		
APR		
MAY		
JUN		
JUL		
AUG		
SEP		
OCT		
NOV		
DEC		

MONTH	DAY	YEAR
JAN		
FEB		
MAR		
APR		
MAY		
JUN		
JUL		
AUG		
SEP		
OCT		
NOV		
DEC		

DO NOT WRITE IN THIS AREA   0520   ©TDS Limited

NCS Trans-Optic MP08-18498-321   A8804   PRINTED IN U.S.A.

7. Are you the primary source of income in your household?
   - ○ Yes
   - ○ No

8. Employment Status
   - ○ Full Time
   - ○ Part Time
   - ○ Temporary Full Time
   - ○ Temporary Part Time

9. On the average, how many hours a day are you in the building?
   - ○ 1 to 2
   - ○ 3 to 4
   - ○ 5 to 6
   - ○ 7 to 8
   - ○ 9 to 10
   - ○ Over 10

10. What time do you usually arrive at the building?
    - ○ 8 a.m. or earlier
    - ○ Between 8 and 9 a.m.
    - ○ Between 9 and 10 a.m.
    - ○ 10 a.m. or later

11. What time do you usually leave the building?
    - ○ 4 p.m. or earlier
    - ○ Between 4 and 5 p.m.
    - ○ Between 5 and 6 p.m.
    - ○ 6 p.m. or later

12. How many hours a week do you usually work in the building?
    - ○ Under 25
    - ○ 25 to 30
    - ○ 31 to 40
    - ○ Over 40

13. How many days have you been absent from work due to illness during the last six months?
    - ○ None
    - ○ 1 to 2
    - ○ 3 to 5
    - ○ 6 to 8
    - ○ 9 to 12
    - ○ Over 12

14. In the past six months, how many days have you left work earlier than you had planned due to feeling unwell?
    - ○ None
    - ○ 1 to 2
    - ○ 3 to 5
    - ○ 6 to 8
    - ○ 9 to 12
    - ○ Over 12

PLEASE RATE THE FOLLOWING QUESTIONS

	NEVER	RARELY	SOMETIMES	ALWAYS
15. In your job, are you able to make decisions on your own?	○	○	○	○
16. Are you free to organize how you do your job?	○	○	○	○
17. Can you control the speed at which you work?	○	○	○	○
18. Are you able to participate in decisions that affect your job?	○	○	○	○
19. Does your job require you to work very fast?	○	○	○	○
20. Does your job require a great deal of concentration?	○	○	○	○
21. Is your job boring or monotonous?	○	○	○	○
22. Is your job stressful?	○	○	○	○
23. Is your job satisfying?	○	○	○	○
24. Does your job require extensive physical exertion?	○	○	○	○
25. Does your job require you to use awkward movements or poor posture?	○	○	○	○
26. Are co-workers helpful in getting your job done?	○	○	○	○
27. Do you find others frustrate your efforts to get your job done?	○	○	○	○
28. Are there enough staff in your department to cope with the work load?	○	○	○	○
29. Is your work area too crowded?	○	○	○	○
30. Do you remain in one location at work each day?	○	○	○	○
31. Does your job take you out of the building?	○	○	○	○
32. Do you leave the building at lunch time?	○	○	○	○
33. Do you find the building claustrophobic?	○	○	○	○

In your primary work area (that in which you spend the most time), how often do each of the following conditions occur?

	NEVER	RARELY	SOMETIMES	ALWAYS
34. Too little air movement	○	○	○	○
35. Too much air movement	○	○	○	○
36. Just the right air movement	○	○	○	○
37. Air too dry	○	○	○	○
38. Air too moist	○	○	○	○
39. Humidity just right	○	○	○	○
40. Air too smokey	○	○	○	○
41. Air too stuffy	○	○	○	○
42. Unpleasant odours in the air	○	○	○	○
43. Temperature too hot	○	○	○	○
44. Temperature too cold	○	○	○	○
45. Temperature just right	○	○	○	○
46. Lighting too bright	○	○	○	○
47. Lighting too dim	○	○	○	○
48. Too much glare on work surface	○	○	○	○
49. Lighting just right	○	○	○	○
50. Too noisy	○	○	○	○
51. Too quiet	○	○	○	○
52. Noise level just right	○	○	○	○

USE A PENCIL FOR ALL ENTRIES

©TDS Limited

How often do you use the following types of equipment at work?

NEVER / RARELY / SOMETIMES / ALWAYS

53. Dictating machine .... ○ ○ ○ ○
54. Typewriter .... ○ ○ ○ ○
55. VDU/Word Processor .... ○ ○ ○ ○
56. Photocopier .... ○ ○ ○ ○
57. Duplicator .... ○ ○ ○ ○
58. Adding machine or calculator .... ○ ○ ○ ○
59. Microfiche .... ○ ○ ○ ○

If you work on a VDU/Word Processor please answer the following questions. If not, please skip this section and continue with Question 74.

How often are the following problems bothersome?

NEVER / RARELY / SOMETIMES / ALWAYS

60. Flickering of the screen .... ○ ○ ○ ○
61. Brightness of the screen .... ○ ○ ○ ○
62. Contrast of the screen .... ○ ○ ○ ○
63. Distance to the screen .... ○ ○ ○ ○
64. Angle of the screen .... ○ ○ ○ ○
65. Glare from lights above or behind the screen .... ○ ○ ○ ○
66. Size of the lettering .... ○ ○ ○ ○
67. Brightness of the lettering .... ○ ○ ○ ○
68. Clarity of the lettering .... ○ ○ ○ ○
69. Distance to the keyboard .... ○ ○ ○ ○
70. Angle of the keyboard .... ○ ○ ○ ○
71. Height of the desk .... ○ ○ ○ ○
72. Seating .... ○ ○ ○ ○

73. Are the keyboard and the screen separate units on the VDU/Word Processer at which you work?
○ Yes   ○ No

ALL RESPONDENTS please continue with Question 74

74. Is your work area
○ Enclosed by walls
○ Enclosed by ceiling height partitions
○ Enclosed by screens
○ Partially enclosed by screens
○ Not enclosed

75. Do you find your chair comfortable?
○ Yes   ○ No   ○ Occasionally uncomfortable

76. Can you see a window from where you normally sit at work?
○ Yes   ○ No

77. Do the windows open?
○ Yes   ○ No   ○ Yes, but not allowed

78. Approximately what distance are you from a window?
○ Less than 5 feet
○ 5 to 10 feet
○ 11 to 20 feet
○ Over 20 feet
○ No windows in the vicinity

How is your work area lit? (Choose all that apply)
79. ○ Fluorescent ceiling light
80. ○ Fluorescent table light
81. ○ Incandescent ceiling light
82. ○ Incandescent table light
83. ○ Natural window light

Are you able to control the following? (Choose all that apply)
84. ○ Work area temperature
85. ○ Work area ventilation
86. ○ Work area lighting
87. ○ Work area humidity

In the past year have any of the following been brought into your work area?
88. ○ Portable heater
89. ○ Desk top fan
90. ○ Portable air cleaner
91. ○ Portable humidifier
92. ○ Negative ion generator
93. ○ Radio/Piped music

94. Do you smoke cigarettes?
○ Yes   ○ No

95. Do you smoke cigars?
○ Yes   ○ No

96. Do you smoke a pipe?
○ Yes   ○ No

97. Do you smoke at work?
○ Yes   ○ No

If you answered YES to Questions 94, 95, 96 or 97, please continue responding to the following questions. If you do not smoke, please continue with Question 101.

98. How many cigarettes do you smoke each day, on average?
○ None
○ 1 or less
○ 2 to 10
○ 11 to 20
○ 21 to 40
○ Over 40

99. How many cigars do you smoke each day, on average?
○ None
○ 1 or less
○ 2 to 4
○ Over 4

©TDS Limited

DO NOT WRITE IN THIS SPACE

100. How many pipesful do you smoke each day, on average?
- ○ None
- ○ 1 or less
- ○ 2 to 5
- ○ Over 5

ALL RESPONDENTS please continue with Question 101.

101. Do you wear corrective lenses at work?
- ○ No
- ○ Reading glasses
- ○ Regular glasses
- ○ Bifocals
- ○ Contact lenses

102. How many cups of coffee and/or tea do you drink, on average, each day at work?
- ○ None
- ○ 1 or less
- ○ 2 to 3
- ○ 4 to 5
- ○ 6 or more

103. Do you drink alcohol?
- ○ Yes
- ○ No

104. How many work days a week do you drink alcohol at lunchtime?
- ○ None
- ○ One
- ○ 2 to 3
- ○ 4 to 5

105. Do you work near people who smoke?
- ○ Yes
- ○ No

Which of the following cooking appliances do you use in your home? (Choose all that apply)
- 106. ○ Gas stove
- 107. ○ Electric stove
- 108. ○ Mixed gas and electric stove
- 109. ○ Microwave oven
- 110. ○ Other

What types of fuel are used to heat your home? (Choose all that apply)
- 111. ○ Gas
- 112. ○ Electricity
- 113. ○ Oil
- 114. ○ Coal
- 115. ○ Wood
- 116. ○ Kerosene/Paraffin
- 117. ○ Other

What types of heating systems are used in your home? (Choose all that apply)
- 118. ○ Forced air
- 119. ○ Radiators
- 120. ○ Fireplace
- 121. ○ Portable heater
- 122. ○ Stove

123. Is your home air conditioned?
- ○ Yes
- ○ No

Have you experienced any of the symptoms listed below while at work in the building? (Please rate each of the following)

	NEVER	RARELY	SOMETIMES	ALWAYS
124. Headache	○	○	○	○
125. Fever	○	○	○	○
126. Dizziness	○	○	○	○
127. Fatigue	○	○	○	○
128. Sleepiness	○	○	○	○
129. Weakness	○	○	○	○
130. Nausea	○	○	○	○
131. Respiratory problems (breathlessness, wheezing)	○	○	○	○
132. Muscular aches of arms, hands or wrists	○	○	○	○
133. Chest pain or tightness	○	○	○	○
134. Backache	○	○	○	○
135. Neckache	○	○	○	○
136. Eye irritation	○	○	○	○
137. Trouble focussing eyes	○	○	○	○
138. Sore or irritated throat	○	○	○	○
139. Nose irritation (itching or running)	○	○	○	○
140. Cold/Flu symptoms	○	○	○	○
141. Depression	○	○	○	○
142. Difficulty concentrating	○	○	○	○
143. Tension or nervousness	○	○	○	○
144. Skin dryness, rash or itching	○	○	○	○
145. Cold extremities (feet, hands, etc.)	○	○	○	○

146. Are you currently taking any courses of prescribed medication?
- ○ Yes
- ○ No

Do you suffer from any of the following?

	YES	NO
147. Migraine	○	○
148. Asthma	○	○
149. Eczema	○	○
150. Hayfever or other allergies	○	○
151. Dysmenorrhea (painful menstruation)	○	○

THANK YOU FOR YOUR ASSISTANCE IN THIS SURVEY

○○○■○○○○○■○○○○○○○○○○○○○○○ 0520

DO NOT WRITE IN THIS AREA

©TDS Limited

References

[1] McDonald, J. C., "Investigation of Employee Health Complaints at Les Terrasses de la Chaudiere," Final Report to TB/PSAC Steering Committee, Treasury Board of Canada, Contract TB/CRTREQ B8059, 1984.
[2] Collett, C. W., Weinkam, J. J., Steeves, J. F., Sterling, E. M., and McIntyre, E. D., "The Building Performance Database: An Analytical Tool for Indoor Air Quality Research," *Postor Presentations,* Indoor Air Quality '86: Managing Indoor Air for Health and Energy Conservation, American Society of Heating, Refrigerating and Air Conditioning Engineers Conference, Atlanta, 1986.
[3] *The Building Performance Database User's Guide,* Theodor D. Sterling Limited, Vancouver, B.C. 1984, available on request.
[4] Sterling, E. M., McIntyre, E. D., Collett, C. W., Meredith, J., and Sterling, T. D., *Environmental Health Review,* Vol. 29, No. 3, 1985, pp. 11–19.
[5] Sterling, E. M. and Sterling, T. D. in *Proceedings,* 5th AIC Conference on the Implementation and Effectiveness of Air Infiltration Standards in Buildings, Air Infiltration Center, Reno, 1–4 Oct. 1984, pp. 17.1–17.13.
[6] Hedge, A., "Building Illness Indices Based on Questionnaire Responses," *Proceedings,* Indoor Air Quality '86: Managing Indoor Air for Health and Energy Conservation, American Society of Heating, Refrigerating and Air Conditioning Engineers Conference, Atlanta, GA, 1986, in press.
[7] *Building Performance Assessment of the Robert Ker Building, Results of Phase 1: Work Environment Survey,* Theodor D. Sterling Ltd., Vancouver, B.C., 1985.
[8] Sterling, E. M. and Sterling, T. D. in *Proceedings,* AIA Research and Design '85, American Institute of Architects, Los Angeles, 14–18 March 1985.
[9] Sterling, E. M. and Sterling, T. D. in *Transactions,* American Society of Heating, Refrigerating and Air Conditioning Engineers Conference, Vol. 91, No. 4, 1985, pp. 520–529.
[10] Rajhans, G. S., *Occupational Health in Ontario,* Vol. 4, No. 4, 1983, pp. 160–167.
[11] Melius, J., Wallingford, K., Keenleyside, R., and Carpenter, J., "Indoor Air Quality—The NIOSH Experience," Annual Meeting of the American Congress of Government Hygienests, Atlanta, 1984.
[12] Kirkbride, J., *Sick Building Syndrome: Causes and Effects,* Health and Welfare Canada, Ottawa, 1985.

Aharon Fradkin[1]

Sampling of Microbiological Contaminants in Indoor Air

REFERENCE: Fradkin, A., **"Sampling of Microbiological Contaminants in Indoor Air,"** *Sampling and Calibration for Atmospheric Measurements, ASTM STP 957*, J. K. Taylor, Eds., American Society for Testing and Materials, Philadelphia, 1987, pp. 66–77.

ABSTRACT: Strategies for sampling airborne biogenic contaminants including fungi, bacteria, viruses, and protozoa should include specific considerations for each group and very often for different species within these groups. The determination of target pollutants and sampling strategy is based on preliminary investigation of the indoor environment by a walk-through inspection and clinical, epidemiological, and immunological features of health effects. The two general approaches are sampling for the detection of a particular pollutant and sampling for the detection of deviations from normal concentrations or specific composition of bioaerosols or both. The principal sampling techniques are based on sedimentation, filtration, impingement, precipitation, centrifugal separation, and impaction. Two samplers, namely, the six-stage Andersen and the all-glass impinger Model AGI-30 have been suggested as standards. General criteria for evaluation of sampling techniques are collection efficiency, sensitivity, reliability, ease of sterilization, maintenance and operation, and effect on viability. Particular considerations for indoor air are sampler noise, particle size discrimination, capacity, and air flow rate. Attention should also be given to sampling location, number of samples per interior, reference data, and methods to confirm the findings of air sampling.

KEY WORDS: air quality, calibration, sampling, atmospheric measurements, airborne microorganisms, air sampling, indoor air pollution, bioaerosols, aeromicroflora, airborne bacteria, airborne fungi, airborne allergins, airborne infections, microbial pollutants, sampling biological aerosols

The growing concern of the public and health authorities about indoor air quality (IAQ) has raised the need for sampling methods for the assessment of air pollution in residential and commercial buildings. Airborne microorganisms have frequently been placed at the bottom of indoor pollutants lists [1,2] and have been given relatively little attention in IAQ studies [3,4]. Nevertheless, some of the best documented cases of indoor-related illnesses have been the result of microbial air pollution [5]. These include contagious diseases that are transferred from person to person, airborne infections from environmental sources, and allergies.

The sampling of indoor air pollutants refers generally to the collection and analysis of airborne agents to determine whether they are associated with adverse health effects or if they occur indoors at higher than normal concentrations. Normal concentrations, in that respect, are those that occur in the outdoor air of the same area. Although there are numerous methods available for the collection of airborne microorganisms and for the assessment of their concentrations, the sampling of microbial contaminants is complex. The indoor air normally contains varying concentrations of heterogenous bioaerosols that have

---

[1] Project scientist, Concord Scientific Corporation, Downsview, Ont., Canada M3H 2V2.

no measurable effect on occupants, and very frequently the predominant species are from outdoor sources such as soil, plants etc. and occur in the outdoor air at higher concentrations [6].

Effective sampling of contaminants, therefore, is dependent on previous information regarding the organisms involved including possible sources, ecological characteristics or effects, and on the application of suitable procedures for their recovery.

This paper is concerned with three aspects of sampling microbiological contaminants in the air, namely, (1) determination of target pollutants, (2) collection, and (3) analysis. These issues are discussed from a practical viewpoint with specific considerations for different indoor situations.

Determination of Target Pollutants

The major microbial pollutants include species of four taxonomically different groups, namely, fungi, bacteria, viruses, and protozoa. Their involvement with indoor air pollution is usually associated with growth in a particular habitat within or in the proximity of the interior and with well-defined health effects. These aspects can be studied by a walk-through inspection and by epidemiological, clinical, and immunological studies. Although they are beyond the scope of this review, these issues are important to air sampling and are discussed briefly.

Fungi

Fungi occur in the air as spores and hyphal fragments, the inhalation of which has been associated with hypersensitivity diseases and infections [5,7]. Although many fungal species are known as allergens, those that belong to *Penicillium* and *Aspergillus* have been given special attention in indoor situations [8]. These have often involved the growth of the fungi on household and building materials, usually where chronic humidity persists [8] and in humidifiers and heating ventilation and air conditioning (HVAC) units [5]. The latter should be inspected carefully because of the effective aerosolization of contaminants through these systems.

Bacteria

The best known bacterial air pollutant is *Legionella pneumophila*, the causative agent of Legionnaires' disease and Pontiac fever [9,10]. Legionella outbreaks have been caused by aerosols from HVAC systems, cooling towers, and showerheads [10]. Since this pathogen is capable of growth in a wide range of natural habitats, potential sources such as stagnant water on rooftops and soil in the proximity of buildings should be taken into consideration [10]. Aerosols of spores of a different type of bacteria, namely, thermophilic actinomycetes, have been generated following growth in humidifiers and HVAC systems. Several species identified, including *Micropolyspora faeni*, *Thermoactinomyces vulgaris*, and *T. candidus*, have been incriminated in cases of hypersensitivity pneumonitis [7]. Most bacterial aerosols in indoor environments are believed to be generated from human sources with gram-positive cocci being the most prevalent [11].

Viruses and Other Contagious Organisms

Riley [12] has considered contagious organisms such as *Mycobacterium tuberculosis* and respiratory viruses including measles and influenza as indoor air pollutants because they are contracted via the air and their transmission is especially efficient indoors. The source of these pathogens is the respiratory tract of infected individuals who discharge the aerosols through sneezing and coughing. This type of air pollution has been especially important in situations where large numbers of individuals are crowded for extended time periods, for example, in schools, military barracks, and hospitals [12].

Protozoa

Naegleria gruberi and possibly other species of amoebae have been associated with "humidifier fever" through contamination of HVAC systems [7]. Although the involvement of airborne amoebae with building outbreaks has been demonstrated by air sampling [13], the extent of their role as pollutants is poorly understood because most studies have not utilized the media required for their isolation.

Evidence from Health Effects

Features of health effects that are associated with indoor environments have been studied by questionnaires, medical check-ups, pulmonary function tests before and after exposure to the air in the building, and immunological methods such as skin tests, Radioallergosorbent Test (RAST), and gel diffusion [7]. These procedures may allow the characterization of target allergens, the detection of their sources, and the elimination of other possible causes. *Legionella* and other infectious microorganisms are best characterized by symptoms and immunological tests [14]. The identification of these pathogens in specimens taken from sick individuals can serve for confirmation.

Collection

In the last 50 years, there have been numerous developments of air samplers for the collection of airborne microorganisms. The literature concerning the different types of samplers and their relative efficiencies has been reviewed [15–18]. Much of this information is based on experimental studies in aerosol chambers and no attempt will be made here to review in depth the voluminous material. To provide the background for the selection of samplers for IAQ investigations, however, a brief overview of the methods available is in order.

The procedures used for sampling of bioaerosols may be classified into gravitational techniques where the collection is based on the settling properties of airborne particles and volumetric methods that are based on active removal of particles from known volumes of air. In addition, the occurrence of pathogens in the air has been demonstrated using laboratory animals.

In most instances, the recovery of microorganisms from the air is dependent on their capacity to grow and multiply after collection [16]. This is the major difference between bioaerosols and other airborne particles. The principal requirement of many of the sampling procedures is, therefore, to retain the viability of the collected agents and to allow for their subsequent cultivation. Excluded are some fungal spores with characteristic morphological features that may be detected without further culturing [19].

Gravitational Methods

According to present concepts of air pollution, gravitational methods are not acceptable for indoor air quality assessment. Nevertheless, these methods have been used widely until as recent as 1983 [20] and therefore should be reviewed.

The most popular gravitational method for the collection of airborne microorganisms involves the exposure of Petri plates of semi-solid (agar) media containing nutrients. Bioaerosols are allowed to settle on the agar surface for a period of one to several hours after which the plates are covered and incubated. The advantages of the settling plates are that they are inexpensive, simple to operate, noiseless [15], and they have been used in many studies, thus providing an abundance of referenced results.

On the other hand, gravitational techniques have several drawbacks that would render them unsuitable for any serious assessment of the degree of contamination of the air. Bourdillon and coworkers [21] have shown that sedimentation collected large particles 200 times more effectively than small particles. The discriminating effect of the particle size can be explained by the equation of the terminal settling velocity (V_t) of aerosol particles,

$$V_t = \frac{P_p D_p^2 g}{18\eta}$$

which is a function of the density of the particle (P_p), the particle diameter (D_p), gravity acceleration (g), and air velocity (η). The particle size is directly proportional to the settling velocity. Another factor that can affect the sedimentation of particles is the air velocity. Investigations of office buildings[2] have shown that the microbial numbers (fungi and bacteria) were 10 to 20 times higher outdoors, in the wind, than inside the buildings where the air was almost calm. This may be explained by the larger volume of air that was in contact with agar surface under wind conditions, compared with the calm air. A volumetric sampler that was run in parallel has found that the concentrations indoors were comparable to the ambient air.

Other gravitational techniques available utilize surfaces coated with adhesives and are aimed at the collection of airborne particles for microscopic examination [22]. One such device, namely, the gravity slide Durham sampler, was recommended as a standard pollen sampler by the American Academy of Allergy [22]. At present, however, the Mold and Pollen Committee is in the process of changing this to the Rotorod, which is a volumetric sampler.[3]

Volumetric Samplings

Based on the principle of quantitative particle removal from the air stream, the volumetric techniques can be classified into filtration, electrostatic precipitation, centrifugal separation, impingement, and impaction. Each technique has special characteristics that can make it suitable or unsuitable for the sampling of a particular situation.

Filtration—The suction of an air stream by a vacuum pump, through filters (usually of 0.45 μm pore size) would result in deposition of particles on the filter surface. Filter samplers are primarily applicable for the collection of hardy spores and less suitable for vegetative cells and viruses that are generally susceptible to dessication. The sampling time recommended is not to exceed several minutes for bacteria and viruses, and longer for spores at flow rates of 5 to 50 L/min [18]. There are various kinds of filters available for air sampling of which cotton, gelatin, and membrane filters have been used most frequently in aerobiological investigations [23]. Comparison between filters for various applications are available in a recent review [15]. Filtration air samplers have several advantages. They are highly

---

[2] A. Fradkin, unpublished data.
[3] Dr. H. Burge, personal communication.

efficient for the recovery of spores and are suitable for a wide range of bioaerosol concentrations. In addition, these samplers are simple to operate and relatively inexpensive, especially if considering that they may be used for sampling of other airborne agents. Another advantage is that the collected material may be analyzed in various ways (see following sections). The need to further process sampled material before incubation, however, may sometimes be considered a drawback because it is laborious and time-consuming. To provide meaningful results, the filtration should be compared with the Andersen sampling and, in any case, where microbes are eluted from a filter, adequate counts in the serial dilution plates should be obtained.[4]

Electrostatic Precipitation—Air samplers based on the concepts of electrostatic precipitation have been designed and applied for the detection of very low concentrations of airborne microorganisms, particularly bacteria and viruses [15]. The recovery of these particles from the air is dictated by the precipitating force (F_s) that is a function of the electric field intensity (E_p) and the particle charge (Q) according to $F_s = E_p \cdot Q$. Most electrostatic precipitators available operate at high sampling rates (500 to 10 000 L/min), which would suggest they may be unsuitable for most indoor applications, particularly if the detection of sources is concerned. Although these samplers have been used mostly as research tools [15], they may be potentially effective for the collection of contagions in indoor environments.

Centrifugal Separation—The centrifugal separation is based on the principle that particles suspended in a rotating air stream are forced to move outwards from the circular path. The formula of the centrifugal force (F_c) is

$$F_c = \frac{mv^2}{R}$$

where m = mass of particle, v = tangential air velocity, and R = radius of curvature. Depending on whether the centrifugal force is generated by rotating the body of the sampler or by the design of a static sampler, these devices have been designated centrifuges and cyclones, respectively. Fannin and Vana [15] have ranked glass cyclone scrubbers as the most suitable overall for collecting viable microorganisms in ambient air. The collection is into a liquid. There are several developments of these samplers that operate at sampling rates from 1.6 L/min to 1 m³/min [15]. Similar to the precipitators, the high volume cyclones may be effective for sampling contagions, although Simard and co-workers [11] have failed to isolate airborne viruses in a building using such a sampler at a flow rate of 950 L/min. The potential drawback of glass cyclone scrubbers is that their operation may be too complex for industrial hygienists who conduct routine IAQ studies.

Unlike cyclones, there are only a few versions of centrifugal samplers and they are by and large inadequate for airborne microorganisms [15].

Of major interest is a relatively new device that is available under the commercial name Biotest RCS [24]. It has been found to detect higher numbers of bioaerosols compared with a slit-to-agar sampler or an impinger [24], and its small dimensions would make it attractive for IAQ studies. Furthermore, the collection onto agar media and the fixed flow rate of 40 L/min are convenient characteristics for many indoor applications. However, any recommendations should await further evaluation. The apparent drawbacks are that this sampler

---

[4] Dr. P. Morey, personal communication.

uses special agar strips that can be obtained only from its manufacturer, and it is not effective for the collection of particles less than 5 μm in diameter.

Impingement—When an air stream is impinged into a liquid medium, the airborne particles are retained in the liquid separated from the gaseous phase. This principle has served for the development of various sampling devices, one of which, namely, the all-glass impinger (AGI-30), has been suggested as a standard sampler for bioaerosols [25]. Using different volumes of collecting liquid and different flow rates, impingers may be adapted for a wide range of microbial concentrations. The AGI-30 is usually operated at 12.5 L/min using a 20-ml sampling liquid [16]. The AGI samplers have been studied in comparison to many other samplers [15], and their major advantages are simple and easy operation, low cost, high efficiency, and reliability. On the other hand, the AGI-30 is not adequate for sampling low concentrations of bioaerosols because the sampling time is restricted by possible evaporation of the sampling fluid. Organisms collected by impingement should be further processed, such as by filtration or plating out on agar media. This can be advantageous if sampling unknown bioaerosols, because more than one type of medium and analysis procedure can be applied for each sample. However, the procedure is less convenient than sampling directly onto agar.

Impaction—The collection of airborne particles by intercepting the air stream by nutrient media is the principle of many of the air samplers presently used in IAQ studies. Particles are impacted on the media surface when their inertia overcome the aerodynamic drag of the air stream. This is a function of several factors including particle size, density, and air velocity. The most commonly used sampler in this category is the Andersen cascade impactor. There are several models of Andersen samplers of which the six-stage has been considered a standard reference sampling instrument for aerobiological studies [18]. This sampler is designed to separate the particles according to six size ranges using stages that supposedly simulate the human respiratory system [26]. A two-stage model of Andersen that discriminates between respirable and non-respirable bioaerosols is also available. However, the latter has been found to yield lower values than the six-stage model when the concentrations of particles smaller than 1 μm in diameter in the air were less than 1000 particles/m³ [27]. Although size discrimination capacity has been considered important for the assessment of human exposure [18], the Andersen sampler has often been used without reference to size distribution of bioaerosols. This may suggest that the use of a modified Andersen that includes only one stage and provides comparable results of total concentrations is sufficient [28]. The Andersen is usually operated at a flow rate of 28.3 L/min for 1 to 20 min, depending on the anticipated concentrations.

The extensive utilization of the Andersen in indoor studies has yielded a good deal of information that can serve as a reference for future studies. Other advantages of this sampler are that it is suitable for different organisms, and the collection is directly onto culture media contained in Petri plates. This, however, involves separate sampling for each type of medium. Another drawback of the Andersen is that it is relatively expensive. Drying of the agar during prolonged sampling is also a potential problem of collection onto agar, but it can be solved by application of a monolayer of oxyethylene docosanol to the agar surface [29].

Another type of impactor, namely, slit-to-agar [21], is commonly used in commercial laboratories and industry at a sampling rate of 28.3 L/min for 1 to 60 min. It does not provide size discrimination, but because the particles are impacted on slowly rotating agar, a relationship between time and microbial aerosol concentration can be established [15]. A recent study comparing the slit-sampler with another impaction sampler, namely, Surface

Air System (SAS), which operates at high flow rates (180 L/min), has found the efficiency of the SAS to be 70 to 80% of that of the former [30]. This apparently was due to low collection efficiency of very small particles (<4 μm). The high sampling rates of the SAS, however, make it suitable for air assessment where low microbial concentrations are expected.

Rotorod is another type of air sampler based on impaction. Unlike the samples just described, this sampler was designed for the collection of spores and pollen for direct microscopic identification. Comparative studies have found the Rotorod to be inferior to a spore trap sampler in ambient air [31], but this sampler was reported to be highly useful in studies of airborne fungi related to mold allergies [19]. The disadvantage of the Rotorod is that it is not effective for particles smaller than 10 to 20 μm diameter size range. Recently, a new impactor, namely, Personal Volumetric Air Sampler (PAS), has been developed by Burkard,[5] which is a small version of the Hirst spore trap, designed for indoor sampling. The PAS is similar to the Rotorod in that particles are collected on a glass slide and can be examined microscopically. According to the manufacturer, the efficiency of the orifice is in the order of 97% for particles in the 5-μm range, which would make the PAS more effective for this range of bioaerosols than the Rotorod.

Collection Media

Bioaerosols may be collected into various media depending on the sampling method used. Organisms that are sampled into liquid or on filters must be transferred to nutrient media for culturing. These procedures are more laborious compared to the collection onto agar, but they may allow high flexibility with regard to sampling duration and the manipulation of the collected material. For example, the microorganisms collected may be, as necessary, concentrated or diluted and samples may be plated out on numerous types of culture media. Furthermore, this material may be examined microscopically or be used for other tests in parallel to the culturing. Direct collection onto the culture media is more convenient than the preceding, but the range of species sampled is limited by sampling time and number of samplers utilized. The use of more than one culture medium is essential for both the sampling of a wide variety of bioaerosols and for identification. This aspect has been well demonstrated by Kozak and co-workers [32] who have shown that *Torula herbarium* failed to sporulate on the collection medium (Sabouraud), but when the fungus was transferred to Moyer's multiple media, sporulation was evident. Formation of spores and their morphology are important features for identification.

Various nutrient media have been tested for the collection of airborne microorganisms. Rose Bengal-Streptomycin agar has been suggested as a selective medium for fungi [33], but Burge et al. [34] have found its performance to be poor compared with potato dextrose agar, Sabouraud's, and malt extract agar. The latter will probably be recommended as standard by the American Conference of Governmental Industrial Hygienists (ACGIH).[6]

Bacteria have been collected onto blood agar, chocolate agar [11], and Trypticase soy agar [35] and viruses into MEM fluid medium supplemented with 0.5% bovine serum albumin, L-glutamine, and gentamin, from which they were transferred to tissue culture for cultivation [11].

Analysis

The distinction between collection and analysis is somewhat artificial because many of the collection considerations dictate which organisms would be recovered. In this context,

---

[5] Rickmansworth, Hertfordshire, WD3 IPJ, U.K.
[6] Dr. P. Morey, personal communication.

analysis is regarded as the identification and quantification of the material collected. The identification of microorganisms requires a high degree of expertise in the different fields including mycology, bacteriology, virology, etc., and no attempt will be made to review the matter. Keeping in mind that this paper is intended to serve professionals with limited training in microbiology, however, some of the principles and the common problems of the analysis are described.

Collected aerosols of microorganisms may be classified into two categories based on their capacity to form colonies on nutrient media. Most sampling procedures are aimed at culturable contaminants. However, there is no relationship between culturability of organisms and their significance as indoor air pollutants. Microorganisms fail to grow when they are dead, dormant, inhibited, or if they are not provided with adequate conditions. These characteristics do not necessarily reflect the capacity of microbial aerosols to cause adverse health effects when inhaled. The detection of non-culturable microbiological pollutants is severely restricted. Fungal spores with distinct morphological features have been identified microscopically after collection on Rotorod and have been recorded as number of spores/ m^3 air [19].

Numbers of culturable bioaerosols are determined after a period of incubation (usually one to several days) under conditions that would correspond to the growth requirements of specific types of microorganisms. For example, mesophiles grow at the range of 25°C, thermophilic actinomycetes should be provided with approximately 55°C, and human pathogens are cultivated at 37°C.

Each colony (or plaque in the case of viruses) that appears on the media surface after incubation is assumed to have developed from a single unit of one or more cells (or viruses) collected in this location. Colony forming unit (cfu), or plaque forming units (pfu) in the case of viruses, is the most commonly used value expressing microbial aerosols. Volumetric procedures usually provide concentrations in volume of air (for example, cfu/m^3). Although this measurement is well imbedded in the literature, it should be stressed that the size of the cfu is dependent on the sampling procedure. For example, in Andersen sampling, airborne particles are continuously impacted on 400 sampling locations on the agar. This may involve the deposition of numerous cells in each location that would be considered one cfu if a colony is developed. In impinger sampling, however, cells are collected into a liquid before being plated out. This may lead to dispersion of cell clusters and subsequently to higher numbers of cfus. The cfus counted after sampling with slit-to-agar or centrifugal samplers are yet different from the preceding. In these cases, the bioaerosols are randomly deposited on the media surface but without further cell dispersion as may occur with impingement. The colonies developed in these procedures represent cell clusters, groups of organisms associated with airborne particles and single cells. Andersen [36] has described a "positive hole correction table" to avoid multiple hits at single impaction sites in his sampler, but the same may not be applicable for the other impactors. Any comparison between cfu results of different samplers is bound to be biased, and there is a clear need for standardization. Dr. P. Morey has suggested that all samplers used for IAQ work should be calibrated or standardized in terms of sampling efficiency against the Andersen, or in instances of very high bioaerosol levels, against the AGI-30.

Sampling Considerations for Indoor Situations

The considerations involved with air sampling of interiors for the detection of microbial pollution may vary from one situation to another and are dependent in general on the findings of the walk-through inspection and features of health effects associated with the interior. Attention should be given to the sampler utilized, sampler conditions, analysis procedures, and sampling protocols.

Selection of Sampler and Sampling Conditions

Although the methods available for sampling of airborne microorganisms have been critically reviewed and two types of air samplers, namely, the six-stage Andersen and the AGI-30, have been recommended as standard devices, there is very limited information concerning the criteria for selection of air samplers for IAQ investigations. In a recent evaluation of air samplers for ambient bioaerosols, Fannin and Vana [15] have used ten criteria, including collection efficiency, sensitivity, reliability, effect on viability, particle size discrimination, cost, etc. These criteria may also be relevant for the assessment of samplers for indoor air. In addition, however, factors such as sampler size, noise, sampling rate, and duration should be considered. Kozak and coworkers [19] have clearly demonstrated that air and source sampling of culturable as well as of non-culturable microflora must be performed in order to detect indoor pollutants. In that case, Andersen and Rotorod have been used for the sampling of airborne molds related to home allergies. The Andersen has also been recommended in other investigations. However, in situations where many samples should be collected in a short time and the portability of the sampler is of importance with regard to the sampling strategy, other samplers may be considered. For example, portable samplers such as SAS or Biotest. There is little information, however, regarding the efficiency and sensitivity of these samplers. The type of organism sought should also be considered in the sampler selection. For routine investigations that are aimed at the detection of overall differences in concentrations between indoor and ambient air, low-volume samples are sufficient. This sampling can be applied in several locations within the interior for the detection of sources of, for example, allergenic molds. Pathogens, on the other hand, can be effective at concentrations of 10^{-12} to 10^{-14} ppb [16] and sampling procedures are applied to provide a "yes or no" answer in cases of outbreaks. For that purpose, high-volume samplers or laboratory animals may be used. Exposure of guinea pigs has been effective for the detection of *Legionella* [14] and *M. tuberculosis* [8], and Artenstein et al. [37] have shown the occurrence of adenovirus aerosols in a military recruit camp by using large-volume electrostatic precipitators. Routine sampling of airborne viruses in buildings using high-volume cyclone-type samplers has shown negative results [11].

Several investigations have reported seasonal and daily variations in concentrations of bioaerosols [5,6]. Solomon [6] has suggested that sampling of indoor molds is much more effective in the winter because during frost-free seasons the ambient mycoflora is abundant and is likely to mask the occurrence of indoor species.

Low sampling rates for short periods may also allow the detection of deviations from the ambient pattern. This is particularly important when sampling onto agar plates because colonies of less prevalent species may be overgrown by commonly occurring fungi. In that case, it may be advantageous to use impingers or other samplers where the collected material can be diluted. Finally, convenience of occupants is sometimes a major consideration for the sampler selection, and it may be suggested that in situations such as hospitals, operating theatres, etc. where a continuous monitoring of the air is involved, attention will be given to this aspect and the less noisy sampler be selected [38].

Choice of Media and Incubation Conditions

Generally, there are two kinds of approaches in investigations of microbiological air pollution, (1) sampling for the detection of a particular pollutant and (2) sampling for the detection of deviations from normal concentrations or specific composition of bioaerosols or both. The selection of media and incubation conditions for each type of investigation are different. When the potential pollutant is well defined by information from a walk-through inspection, source sampling, or health effects, the conditions required specifically for its growth may be provided. For example, cases of *Legionella* may be investigated by using charcoal-yeast extract agar supplemented with ACES buffer and alpha-ketoglutarate that is selective for isolation of *L. pneumophila* [14], and incubation conditions of 37°C. If no previous information regarding the characteristics of the potential pollutants is available, however, the use of more general media, perhaps of different compositions and several incubation temperatures may be recommended. In any case, it should be stressed that the colonies developed are not necessarily of the most prevalent species and that the results of air sampling only reflect the bioaerosols for which the media and incubation conditons would be most suitable.

Protocols for Indoor Air Sampling

Protocols for indoor air sampling address aspects that are not necessarily related to the sampling technique but are of significant importance for the detection of pollutants and their sources. These include the location of the sampler, number of samples per interior, reference information, and methods to confirm the findings of air sampling. Protocols for sampling airborne microorganisms in offices, homes, and other indoor situations are presently being developed by ACGIH [39] and the American Society for Testing and Materials (ASTM) [40], and should be consulted. Briefly, considerations of sampler location should include potential outdoor and indoor sources, and the connection of effects to different rooms within the interior. For example, by comparing the microaeroflora in the proximity of the HVAC systems outlet with that in remote locations, it may be possible to detect whether the systems are the source of pollution. Similar considerations may be applied in situations where only inhabitants at a particular area show health effects [7]. At present, there is no information available to suggest the number of samples required. Most studies have included several samplings in different locations indoors and at least one ambient sampling. However, very little is known about reproducibility of results from sampling at the same location. The lack of exposure guidelines for acceptable concentrations of bioaerosol makes the interpretation of indoor air sampling results complex.

To determine whether the air is contaminated, reference to outdoor concentrations, to results of other sampling procedures, to concentrations of non-culturable bioaerosols, and to species identified in source sampling should be made. In addition, the confirmation of the results may involve immunological assays where antigenic and allergenic extracts of the sampled material will be tested against sources obtained from exposed individuals [7].

In summary, it may be suggested that each study of microbial contamination of indoor air will be based on as much information as possible regarding the specific situation and, where information gaps exist, common sense will be applied.

Acknowledgments

I gratefully acknowledge the support of the Natural Sciences and Engineering Research Council of Canada and of Concord Scientific Corporation (CSC). Special thanks for assistance in the preparation of this paper are due to Dr. R. B. Caton and Ms. M. J. Ghent of CSC.

References

[1] Spengler, J. D. and Sexton, K., *Science,* Vol. 221, 1983, pp. 9–17.
[2] *Indoor Air Quality and Human Health,* I. Turiel, Ed., Stanford University Press, Stanford, CA, 1985.

[3] Nagda, N. L. and Rector, H. E., *Guidelines for Monitoring Indoor Air Quality*, Publication No. EPA 600/4-83-046, U. S. Environmental Protection Agency, Washington, DC, 1983.

[4] *Inventory of Current Indoor Air Quality—Related Research*, Publication No. EPA 600/7-81-119, U. S. Environmental Protection Agency, Washington, DC, 1981.

[5] Solomon, W. R. and Burge, H. A. in *Indoor Air Quality*, P. J. Walsh, C. S. Dundey, and E. D.Copenhaver, Eds., CRC Press, Boca Raton, FL, 1984, pp. 173–191.

[6] Solomon, W. R., *Journal of Allergy and Clinical Immunology*, Vol. 56, 1975, pp. 235–242.

[7] Kreiss, K. and Hodgson, M. J. in *Indoor Air Quality*, P. J. Walsh, C. S. Dundery, and E. D. Copenhaver, Eds., CRC Press, Boca Raton, FL, 1984, pp. 87–106.

[8] *Indoor Pollutants*, National Research Council, National Academy of Science Press, Washington, DC, 1981.

[9] *Legionella, Proceedings*, 2nd International Symposium, C. Thornsberry, A. Balows, J. C. Feely, and W. Jakubowski, Eds., American Society of Microbiology, Washington, DC, 1984.

[10] Imperato, P. J., *Bulletin of the New York Academy of Medicine*, Vol. 57, 1981, pp. 922–935.

[11] Simard, C., Trudel, M., Paquette, G., and Payment, P., *Journal of Hygiene*, Vol. 91, 1983, pp. 277–286.

[12] Riley, R. L., *Environment International*, Vol. 8, 1982, pp. 317–320.

[13] Edwards, J. H., *British Journal of Industrial Medicine*, Vol. 37, 1980, pp. 55–62.

[14] Feeley, J. L. in *Legionella, Proceedings*, 2nd International Symposium, C. Thornsberry, A. Balows, J. C. Feeley, and W. Jakubowski, Eds., American Society of Microbiology, Washington, DC, 1984, pp. 283–284.

[15] Fannin, K. F. and Vana, S. L., *Development and Evaluation of an Ambient Viable Microbial Air Sampler*, Publication No. EPA 600/1-81-069, U. S. Environmental Protection Agency, Cincinnati, 1981.

[16] Chatigny, M. A. in *Occupational Respiratory Disease Report*, Vol. I, J. A. Merchand, Ed., National Institute for Occupational Safety and Health, Centers for Disease Control, U. S. Public Health Service, 1981, pp. 117–163.

[17] *An Introduction to Experimental Aerobiology*, R. L. Dimmick and A.B. Akers, Eds., Wiley, New York, 1969.

[18] Chatigny, M. A. in *Air Sampling Instruments for Evaluation of Atmospheric Contaminants*, 6th ed., American Conference of Governmental Industrial Hygienists, Cincinnati, 1983.

[19] Kozak, P. P., Gallup, J., Cummins, L. H., and Gillman, S. A., *Annals of Allergy*, Vol. 45, 1980, pp. 85–89 and 167–176.

[20] Rogers, S. A., *Annals of Allergy*, Vol. 50, 1983, pp. 37–40.

[21] Bourdillon, R. B., Lidwell, O. M., and Thomas, J. C., *Journal of Hygiene*, Vol. 41, 1941, pp. 197–224.

[22] Solomon, W. R. in *Allergy*, E. Middleton, C. E. Reed, and E. F. Ellis, Eds., The C. V. Mosby Company, St. Louis, 1983, pp. 1143–1190.

[23] Rotter, M., Koller, W., Flamm, H., Resch, W., and Schedling, J. in *Airborne Transmission and Airborne Infection*, 4th International Symposium on Aerobiology, J. F. P. Hers and K. C. Winkler, Eds., Wiley, New York, 1973, pp. 47–50.

[24] Delmore, R. P. and Thompson, W. N., "A Comparison of Air-Sampler Efficiencies," presented at the Annual Meeting of the Society for Industrial Microbiology, Flagstaff, AZ, 10–15 Aug. 1980.

[25] Brachman, P. S., Ehrlich, R., Eichenwald, H. F., Cabelli, V. J., Kethley, T. W., Madin, S. H., Maltman, J. R., Middlebrook, G., Morton, J. D., Silver, I. H., and Wolf, E. K., *Science*, Vol. 144, 1964, p. 1295.

[26] *The Andersen Sampler Brochure*, Bulletin 176–5, Andersen Samplers Incorporated, Atlanta, GA.

[27] Gillespie, V. L., Clark, C. S., Bjornson, H. S., Samuels, S. J., and Holland, J. W., *Journal*, American Industrial Hygiene Association, Vol. 42, 1981, pp. 858–864.

[28] Jones, W., Morring, K., Morey, P., and Sorenson, W. *Journal*, American Industrial Hygiene Association, Vol. 46, 1985, pp. 294–298.

[29] May, K. R., *Applied Microbiology*, Vol. 18, 1969, pp. 513–514.

[30] Lach, V. *Journal of Hospital Infection*, Vol. 6, 1985, pp. 102–107.

[31] Solomon, W. R., Burge, H. A., Boise, J. R., and Becker, M., *International Journal of Biometeorology*, Vol. 24, 1980, pp. 107–116.

[32] Kozak, P. P., Gallup, J., Cummins, L. H., and Gillman, S. A., *Annals of Allergy*, Vol. 43, 1979, pp. 88–94.

[33] Morring, K. L., Sorenson, W. G., and Attfield, M. D., *Journal*, American Industrial Hygiene Association, Vol. 44, 1983, pp. 662–664.

[34] Burge, H. A., Solomon, W. R., and Boise, J. R., *Journal of Allergy and Clinical Immunology*, Vol. 60, 1977, pp. 199–203.

[35] Placencia, A. M., Peeler, J. T., Oxborrow, G. S., and Danielson, J. W., *Applied and Environmental Microbiology,* Vol. 44, 1982, pp. 512–513.
[36] Andersen, A. A., *Journal of Bacteriology,* Vol. 76, 1958, pp. 471–478.
[37] Artenstein, M. S., Miller, S., Rust, H., and Lamson, T. H., *American Journal of Epidemiology,* Vol. 85, 1967, pp. 479–485.
[38] Gould, J. C. "Airborne Pathogenic Bacteria in a Tissue Transplant Unit,"in *Aerobiology, Proceedings of the Third International Symposium,* I. H. Silver, Ed., Academic Press, London, 1970.
[39] Morey, P. R., Hodgson, J. M., Sorenson, W. G., Kullman, G. J., Rhodes, W. W., and Visvesvara, G. S., *Annals,* American Conference of Governmental Industrial Hygienists, Vol. 10, 1984, pp. 21–35.
[40] American Society for Testing and Materials, Indoor Air Subcommittee D-22.05, Task Group on Biological Aerosols D–22.05.06.

Ambient Air

Harry L. Rook[1]

Precision and Accuracy Assessment Derived from Calibration Data

REFERENCE: Rook, H. L., **"Precision and Accuracy Assessment Derived from Calibration Data,"** *Sampling and Calibration for Atmospheric Measurements, ASTM STP 957,* J. K. Taylor, Ed., American Society for Testing and Materials, Philadelphia, 1987, pp. 81–86.

ABSTRACT: During the past ten years, the importance of assessing the precision and the accuracy of data derived from standard methods has become recognized. Both the American Society for Testing and Materials (ASTM) and the U.S. Environmental Protection Agency (EPA) have developed policies requiring precision and accuracy assessment for methods before they are designated as standard methods or reference methods. The difficulty in implementing these policies is not in developing a meaningful uncertainty estimate for a given method, it is in separating the random from the systematic components of the error. In this paper, a method of separating error components into precision and bias is given. The method uses calibration data taken repeatedly at a fixed point using a standard whose total uncertainty is small compared to the uncertainty of the measurement method.

KEY WORDS: calibration, uncertainty estimate, precision, accuracy, data analysis, sampling, air quality, atmospheric measurements

The two most important pieces of information derived from a chemical analysis are the concentration of the analyte of interest and the uncertainty associated with the analysis. These parameters are generally estimated from an analytical data set that is obtained by replicate analysis of a sample and is representative of the system of interest. Then, the true value is usually approximated by the mean of the data set and the uncertainty is usually approximated by the standard deviation of the set. The error, however, has two basic components, the random and the systematic component, which are often hard to separate. It is very important to separate these components, however, because the approach taken to improve a measurement process is often very different depending upon which component of error is dominant. This paper presents a method for the separation of the random and the systematic error components using calibration data where the calibration is carried out using a standard with a small uncertainty.

Procedure

The calibration of an analytical system or instrument is usually carried out with a series of standards where the component of interest is well known and assumed to be correct. The concentration of the component of interest is varied over the working range of the instrument with at least five and usually not more than ten different calibration standards. Then a relationship between the system output or signal and the analyte concentration is constructed.

---

[1] Supervisory research chemist, National Measurement Laboratory, National Bureau of Standards, Center for Analytical Chemistry, Gas and Particulate Science Division, Gaithersburg, MD 20899.

TABLE 1—*Atomic percent of uranium 235 in SRM U005a.*

Sample No.	Atomic %, U-235
C053	0.50739
C021	0.50523
C013	0.50433
C051	0.50733
C052	0.50815
C030	0.50826
C032	0.50606
C064	0.50527
C066	0.50696
C055	0.50741
$\overline{X}$	0.50664
$S_{\overline{x}}$	±0.00134
t	0.50640
W	±0.00136
$\hat{B}$	0.00025
$|\overline{x} - t|$	0.00024

In simple analytical systems, that relationship is often linear over the concentration range of interest, although more complex models are sometimes needed to fit the data. Nonlinear calibration curves usually require larger sets of calibration standards to ensure accurate calibration. An important part of the calibration process is to develop or obtain at least one verification standard that has a known uncertainty in its analyte concentration. With this standard, an estimate of the analytical error can be derived and, more importantly, estimates of method precision and bias can be obtained.

The problem of separating the random from the nonrandom error component can be addressed if either of two conditions are met in the verification standard. The standard must have a "true" value, that is, the analyte concentration has an uncertainty that is very small compared to the error in the measurement process; or the calibration is carried out with two independent verification standards that have a difference, the estimate of the true value that is very small compared to the error in the measurement process. For this procedure to work well, the error in the calibration standard should be at least five times smaller than the expected measurement error. With this condition met for the calibration standard, a single point calibration is repeated, usually five to ten times, and the mean and the estimate of the variances are calculated for the calibration data set.[2]

$$\overline{x} = \frac{1}{n} \sum_{i=1}^{n} x_i \tag{1}$$

$$S^2_{\overline{x}} = \frac{\sum_{i=1}^{n} (x_i - \overline{x})^2}{n - 1} \tag{2}$$

where

x_i = an individual calibration point in the set,
$\overline{x}$ = the mean of the calibration set, and
$S^2_{\overline{x}}$ = the estimate of the measurement variance.

[2] Natrella, M. G., *Experimental Statistics*, National Bureau of Standards Handbook 91, U.S. Government Printing Office, Washington, DC, Aug. 1963.

TABLE 2—*Atomic percent of uranium 235 in SRM U005a.*

Sample No.	Atomic %, U-235
C100	0.51133
C101	0.50941
C029	0.51015
C008	0.50840
C003	0.51363
C061	0.50770
C063	0.50706
C051	0.50892
C059	0.50775
C060	0.50938
$\overline{X}$	0.50937
$S_{\overline{x}}$	±0.00196
t	0.50640
W	±0.00377
$\hat{B}$	0.00313
$|\overline{x} - t|$	0.00297

Using this calibration data set, the standard deviation of the set is an estimate of the random error component of the measurement system and the difference between the mean of the set, $\overline{x}$, and the known or "true" value of the calibration standard, t, is usually used to estimate the measurement bias. The problem with this method of estimating the bias is that the mean value, $\overline{x}$, derived from the data set is based on a small number of observations and has a much greater uncertainty than the estimate of the true value that is usually extensively evaluated. Another method of estimating the bias is to construct a mean squared error (MSE) statistic that contains both the random and the nonrandom components of the error and then separate the error components by subtracting the variances from the MSE.

$$W^2 = \frac{\sum_{i=1}^{n} (x_i - t)^2}{n - 1} \tag{3}$$

where

t = the "true" value of the calibration standard,
x_i = an individual calibration point in the set, and
W^2 = the estimate of the MSE.

TABLE 3—*Manganese in rice flour.*

Sample No.	Concentration of Mn, ppm
0.1	20.49
0.2	20.12
0.3	20.05
0.4	20.54
0.5	20.45
$\overline{X}$	20.32
$S_{\overline{x}}$	±0.24
t	20.10
W	±0.35
$\hat{B}$	0.26
$|\overline{x} - t|$	0.23

TABLE 4—*Manganese in oyster tissue.*

Sample No.	Concentration of Mn, ppm
01	16.80
02	16.51
03	16.96
04	16.10
05	16.62
$\overline{X}$	16.60
$S_{\overline{x}}$	±0.33
t	17.50
W	±1.06
$\hat{B}$	1.01
$|\overline{x} - t|$	0.90

The quantity W^2 is closely related to a statistical estimation for the MSE of a single measurement, x. The statistically defined theoretical MSE of X is $E(x - t)^2$. In terms of the statistical theory of estimation, W^2 is positively biased on an estimate of $E(x - t)^2$ because the expected value of W^2 is $(n/n - 1)E(x - t)^2$.

From the preceding calculation, one can obtain the usual estimate of the random error component using the square root of the sample variance, S^2. This suggests that one might estimate the bias by taking the square root of the difference between the sample variance, S^2, and the estimate of MSE, W^2.

$$\text{Precision} = (S^2)^{1/2} \tag{4}$$

$$\text{Estimate of bias } (\hat{B}) = (W^2 - S^2)^{1/2} \tag{5}$$

By substituting Eqs 2 and 3 into Eq 5 and reducing algebraically, the bias becomes a variation of the familiar $(t - x)$ formulation

$$\hat{B} = \sqrt{\frac{n}{n - 1}} |t - x| \tag{6}$$

TABLE 5—*Atomic percent of uranium 234 in SRM U005a.*

Sample No.	Atomic %, U-234
C053	0.00396
C021	0.00389
C013	0.00406
C051	0.00384
C030	0.00404
C032	0.00406
C064	0.00399
C066	0.00403
C055	0.00402
$\overline{X}$	0.00399
$S_{\overline{x}}$	±0.000074
t	0.00340
W	±0.000629
$\hat{B}$	0.000622
$|\overline{x} - t|$	0.000590

TABLE 6—*Atomic percent of uranium 234 in SRM U0059.*

Sample No.	Atomic %, U-234
C029	0.00367
C059	0.00407
C060	0.00292
C061	0.00361
C101	0.00334
C051	0.00325
C063	0.00383
C100	0.00388
C003	0.00344
C008	0.00317
$\bar{X}$	0.00352
$S_{\bar{x}}$	±0.000358
t	0.00340
W	0.000379
$\hat{B}$	0.000126
$\lvert \bar{x} - t \rvert$	0.000120

Results

The method of separating the random variance from the total variance leads to an estimate of the bias that is always larger than the widely used $\lvert t - \bar{x} \rvert$ method by the factor of $\sqrt{n/(n - 1)}$. This value gives a more useful estimate of the bias, in that, the smaller the data set used to estimate the mean, the larger the uncertainty in the estimate of the bias. The converse is true, in that, when data sets of larger than about 30 observations are used to estimate the mean, the two methods of estimating the bias converge to essentially the same value.

The description method of separating the random from the systematic error components was tried on six data sets containing varying numbers of observations and varying levels of imprecision and bias. The individual data and the estimates of imprecision and the two different estimates of bias are given in Tables 1 through 6. Tables 1 and 2 list data from two different mass spectrometers used to measure the uranium-235 atomic percent in a National Bureau of Standards (NBS) Standard Reference Material (SRM) U-005a. In each case, ten individual observations were made to form the set. In Table 1, the total measurement uncertainty is small and is comprised almost completely of the imprecision, thus both estimates of the bias are essentially the same. In Table 2, the difference between the imprecision and the bias are considerably larger than that of the certified value, t. The bias is significantly larger than the imprecision and therefore will yield the largest improvement to any future data set if corrected. Tables 3 and 4 are examples of trace element data in natural product SRMs. They are examples of calibration data where the imprecision is large relative to the uncertainty of the certified value of the calibration standard. In Table 3, the bias is of the same order as the imprecision. In Table 4, the bias is substantially larger. In both cases, there were only five observations in each data set, thus the technique using $\hat{B}$ yields a more useful estimate of bias. Tables 5 and 6 are again mass spectrometry data but of one of the minor uranium isotopes. In Table 5, the data indicate an unusually large bias resulting from an improper calibration. In Table 6, the data have a large random error component but with a normal level of bias. The data from Table 5 indicate that once the bias is eliminated by recalibration, no additional work on the analytical method would improve the data since the level of imprecision is of the same magnitude as the ion counting statistics of the mass spectrometer. The data from Table 6 are unusual, in that, the imprecision is very much larger than that expected from counting statistics. The bias calculated here is not significant since it is much smaller than the imprecision and the error in the estimate of imprecision will dominate the error in the calculation of the bias no matter which method is used.

Conclusions

The data presented in this paper were chosen to represent those normally encountered relative to imprecision and bias. In all cases, the estimates of bias obtained using $\hat{B}$ were larger than those obtained by the $|x - t|$ method but converged to essentially the same value when the total number of observations in the data set became large. The percent difference between the two methods of computing the bias term became larger as the number of observations in the data set became smaller to the limit of $\sqrt{2}$ as defined by $n = 2$ in Eq 6.

John C. Puzak[1] *and Frank F. McElroy*[1]

The EPA's Role in the Quality Assurance of Ambient Air Pollutant Measurements

REFERENCE: Puzak, J. C. and McElroy, F. F., **"The EPA's Role in the Quality Assurance of Ambient Air Pollutant Measurements,"** *Sampling and Calibration for Atmospheric Measurements, ASTM STP 957,* J. K. Taylor, Ed., American Society for Testing and Materials, Philadelphia, 1987, pp. 87–100.

ABSTRACT: Ensuring the quality and integrity of ambient air pollution measurements encompasses a wide variety of activities and procedures. The U.S. Environmental Protection Agency's (EPA's) roll in the quality assurance of monitoring data is to assist monitoring agencies by providing those quality assurance services or functions that (1) are common to most monitoring agencies and would be uneconomical to duplicate in each agency's program; (2) are too costly for individual monitoring agencies to carry out; (3) need to be uniform across the nation's monitoring agencies to facilitate universal data interpretation, comparison, evaluation, and common analysis; or (4) are more appropriately derived from a source independent from the monitoring agency. These functions are focused in three principle areas: (1) standardized and validated monitoring methods, (2) air monitoring guidance and technical assistance, and (3) assessment of the quality of monitoring data being obtained. These functions are discussed along with overview descriptions and summaries of results of specific EPA quality assurance programs including reference and equivalent methods, the *Quality Assurance Handbook* and other guidance documents, availability of various pollutant concentration standards, and assessment and reporting of the precision and accuracy of air monitoring data.

KEY WORDS: ambient air, air pollutants, sampling, air quality, calibration, atmospheric measurements, quality assurance

Ensuring the quality and integrity of ambient air pollution measurements encompasses a wide variety of activities and procedures. In the various networks of State and Local Air Monitoring Stations (SLAMS), most of these activities are carried out by the state or local agency as a part of their overall data collection process. The role of the U.S. Environmental Protection Agency (EPA) in this scheme is to assist these monitoring agencies by providing those quality assurance services or functions that (1) are common to most monitoring agencies and would be uneconomical to duplicate in each agency's program; (2) are too costly for individual monitoring agencies to carry out; (3) need to be uniform across the nation's monitoring agencies to facilitate universal data interpretation, comparison, evaluation, and common analysis; or (4) are more appropriately derived from a source independent from the monitoring agency.

[1] Deputy director, EMSL, and environmental engineer, respectively, U.S. Environmental Protection Agency/Environmental Monitoring Systems Laboratory (EPA/EMSL), Research Triangle Park, NC 27711.

Pursuant to these goals, EPA's quality assurance support to states' SLAMS monitoring programs is focused in three principle areas, namely:

1. provision of standardized and validated monitoring methods,
2. air monitoring guidance and technical assistance, and
3. assessment of the quality of monitoring data being obtained.

Taken together, these functions provide the states with a strong, fundamental basis and framework to support and augment their overall quality assurance programs. These functions, along with overview descriptions and summaries of the results of specific associated EPA quality assurance support programs, are discussed in the following sections of this paper. Most of these programs have been developed or evolved over the more than 15 years since the National Ambient Air Quality Standards were first promulgated, and are still subject to continual review, evaluation, and further refinement.

Standardized and Validated Monitoring Methods

Reference and Equivalent Methods

Certainly one of the most important and basic elements in obtaining quality ambient air monitoring data is the use of adequate and reliable monitoring methods. A prominent example of EPA's support in this area is the Reference and Equivalent Method program, administered by the Quality Assurance Division of EPA's Environmental Monitoring Systems Laboratory (EMSL) at Research Triangle Park, NC. This program, established in 1975, defines specific performance and other requirements for monitoring methods used by states in their SLAMS compliance monitoring networks and prescribes detailed test procedures by which the methods must be tested and approved.

Under this program, monitoring methods are categorized as either (1) manual methods that are characterized by a detailed, step-by-step, manually executed procedure, or (2) automated methods that are automatic, self-contained analyzers employing an instrumental measurement principle. For each of the criteria pollutants, namely sulfur dioxide (SO_2), ozone (O_3), carbon monoxide (CO), nitrogen dioxide (NO_2), lead (Pb), and total suspended particulate matter (TSP), either a manual reference method (SO_2, Pb, TSP) or a reference measurement principle and calibration procedure (O_3, CO, NO_2) is specified in an Appendix to 40 CFR Part 50 [1] of the EPA monitoring regulations. These Appendixes represent pollutant measurement technology that has been extensively tested and known to be adequate for compliance monitoring purposes. In the former case, the reference method is completely specified in the Part 50 Appendix. However, in the latter case (automated methods), analyzers are designated as reference methods only if they (1) employ the measurement principle and calibration procedure specified in the appropriate appendix of Part 50, and (2) demonstrate that they meet performance specifications and other requirements prescribed in the EPA regulation 40 CFR Part 53 [2]. Hence, while manual reference methods are unique, there can be (and in fact are) multiple reference methods (analyzer models) for O_3, CO, and NO_2.

Manual monitoring methods different than one of the specified reference methods, or automated methods based on a measurement principle and calibration procedure different than those specified for reference methods, may be designated as equivalent methods.[2] For

---

[2] Although there is currently no provision for designating equivalent methods for TSP, proposed amendments to the regulations would provide for designation of both reference and equivalent methods for PM_{10}.

TABLE 1—*General requirements for reference and equivalent methods.*

	Reference Method	Equivalent Method
Manual	None	Consistent Relationship Test (Subpart C)
Automated	1. Specified measurement principle and calibration procedure 2. Laboratory performance tests (Subpart B) 3. No automated reference methods designated if reference method is a manual method.	1. Laboratory performance tests (Subpart B) 2. Consistent Relationship Tests (Subpart C)

use in SLAMS monitoring networks, both reference and equivalent methods are equally acceptable.

Equivalent methods are designated based on their having a "consistent relationship" to a reference method; that is, an equivalent method must demonstrate that it produces measurements equivalent to measurements by a reference method when both methods measure identical samples of actual ambient air. Automated equivalent analyzers must also meet the same performance specifications that are applicable to reference methods. The basic requirements for manual and automated reference and equivalent methods are summarized in Table 1. The actual performance requirements applicable to both reference and equivalent automated methods are listed in Table 2, and the consistent relationship requirements for equivalent methods are given in Table 3. These and the other requirements for designation of reference and equivalent methods (for example, content of the instruction manual, adequacy of operation and calibration procedures, approval of modifications, quality assurance of production analyzers offered for sale, etc.) are set forth formally in 40 CFR Part 53 of the EPA monitoring regulations, along with the detailed test procedures for determining whether candidate methods meet the performance specifications.

TABLE 2—*Performance specifications for automated methods.*

Performance Parameter	Units	SO_2	O_3	CO	NO_2
1. Range	ppm	0 to 0.5	0 to 0.5	0 to 50	0 to 0.5
2. Noise	ppm	0.005	0.005	0.50	0.005
3. Lower detectable limit	ppm	0.01	0.01	1.0	0.01
4. Interference equivalent					
Each interferent	ppm	±0.02	±0.02	±1.0	±0.02
Total interferent	ppm	0.06	0.06	1.5	0.04
5. Zero drift, 12 and 24 h	ppm	±0.02	±0.02	±1.0	±0.02
6. Span drift, 24 h					
80% of upper range limit	%	±20.0	±20.0	±10.0	±20.0
20% of upper range limit	%	±5.0	±5.0	±2.5	±5.0
7. Lag time	min	20	20	10	20
8. Rise time	min	15	15	5	15
9. Fall time	min	15	15	5	15
10. Precision					
20% of upper range limit	ppm	0.01	0.01	0.5	0.02
80% of upper range limit	ppm	0.015	0.01	0.5	0.03

TABLE 3—*Test requirements for the Consistent Relationship Test for equivalent methods.*

| Pollutant | Concentration Range | Simultaneous Measurements Required | | | | Maximum Discrepancy Specification, ppm |
| | | 1 h | | 24 h | | |
		1st set	2nd set	1st set	2nd set	
Ozone	Low 0.06 to 0.10	5	6	...	...	0.02
	Med 0.15 to 0.25	5	6	...	...	·0.03
	High 0.35 to 0.45	4	6	...	...	0.04
	Total	14	18			
Carbon monoxide	Low 7 to 11	5	6	...	...	1.5
	Med 20 to 30	5	6	...	...	2.0
	High 35 to 45	4	6	...	...	3.0
	Total	14	18			
Sulfur dioxide	Low 0.02 to 0.05	...	...	3	·3	0.02
	Med 0.10 to 0.15	...	...	2	3	0.03
	High 0.40 to 0.50	7	8	2	2	0.04
	Total	7	8	7	8	
Nitrogen dioxide	Low 0.02 to 0.08	...	...	3	3	0.02
	Med 0.10 to 0.20	...	...	2	3	0.02
	High 0.25 to 0.35	...	...	2	2	0.03
	Total			7	8	

To date, EPA has designated 60 reference and equivalent methods. These methods, together with the three manual reference methods specified in Part 50, are identified officially in an EPA publication [3]. This list is distributed to the states, regional offices, and other interested monitoring agencies. In addition to processing applications for reference and equivalent methods and maintaining the list of designated methods, the Quality Assurance Division (QAD) also reviews and approves proposed modifications to designated methods by either analyzer manufacturers or users, conducts post-designation tests or other evaluations of designated analyzers, develops improved methods, adapts new measurement principles and calibration procedures for reference methods, and carries out other related activities.

Besides providing high quality monitoring methodology for determining compliance with the ambient air quality regulations, the reference and equivalent methods program has also provided at least two other important benefits.

1. *Flexibility*—As soon as new monitoring techniques are developed and demonstrated to pass the requirements, they can be used immediately for pollutant monitoring. There is no need to change existing regulations to incorporate new or improved monitoring methods.

2. *Reduced Method Development Costs to EPA*—Once the monitoring method requirements were promulgated in the *Code of Federal Regulations*, EPA funds were no longer required to develop monitoring methods for the criteria pollutants. Instrument manufacturers generally have assumed the burden to both develop the methods and demonstrate their acceptability, allowing EPA to apply its resources elsewhere.

Other Method Standardization Activities

The QAD carries out many other method standardization activities for non-criteria pollutants, as well. These activities have recently included developing a new, more sensitive

TABLE 4—Quality Assurance Handbook *volumes*.

Volume	Title	EPA Publication Number
I	Principles of QA	EPA 600/9-76-005
II	Ambient air methods	EPA 600/4-77-027a
III	Stationary source methods	EPA 600/4-77-027b
IV	Meteorological monitoring	EPA 600/4-82-060
V	Precipitation measurements	EPA 600/4-82-042a & b

and accurate method for monitoring ambient concentrations of nonmethane organic compounds, publishing a list of methods for determining toxic organic compounds in ambient air [4], assisting in the standardization of methods for measuring asbestos in the air, and participating in the development of an agencywide guideline for validating environmental monitoring methods of all types.

Guidance and Technical Assistance

Availability of adequate, standardized monitoring methods is of little benefit if those methods are not used correctly and conscientiously by knowledgeable analysts. Consequently, the QAD has long recognized the need for supplemental instruction and guidance to accompany the methods widely used for ambient monitoring. Expertise and experience gained in the development, standardization, and use of monitoring method must be passed on to method users to optimize the quality and usefulness of the data obtained.

Quality Assurance Handbook

One particularly comprehensive guidance document for this purpose is QAD's *Quality Assurance Handbook* [5]. First published in 1976, this detailed document containing both general and specific quality assurance procedures and information has been continually expanded, refined, and updated over the years and has become the "QA bible" among monitoring agencies and other users of the most popular methods for monitoring the criteria pollutants and other environmental parameters.

The *Quality Assurance Handbook* currently consists of five volumes, identified in Table 4. Volume I is devoted to basic, general principles of quality assurance. The subsequent volumes of the *Handbook* cover media-specific general topics and method-specific aspects of many of the most widely used methods. General aspects of quality assurance applicable to ambient air monitoring systems (Section 2.0) and the specific ambient air methods are covered in Volume II. Similarly, general topics and specific methods for stationary source monitoring methods are covered in Volume III. In general, an entire section of the *Handbook* is devoted to each of the methods covered. The comprehensive scope of each of these method-specific sections is indicated by the list of topics for a typical method-specific section shown in Table 5.

Other Guidance and Technical Assistance

Over the years, QAD has issued a great deal of technical assistance to the user community in the form of technical reports, instruction manuals, status reports, guidance documents, seminars, symposia, workshops, and telephone assistance. Of particular interest is a series of comprehensive technical publications called Technical Assistance Documents (TADs).

TABLE 5—*Outline of a typical method-specific section.*

1. Procurement of equipment and supplies
2. Calibration of equipment
3. Operation and procedure
4. Data reduction, validation and reporting
5. Maintenance
6. Auditing procedure
7. Assessment of monitoring data for precision and accuracy
8. Recommended standards for establishing traceability
9. Complete method description
10. References
11. Data forms

The currently available TADs are identified in Table 6. Each of these TADs contains an in-depth discussion of specific and particularly complex areas of ambient monitoring and provides detailed guidance and instruction, and, where appropriate, specific procedures needed for effective quality assurance in these areas. These TADs are available from the National Technical Information Service.[3]

Pollutant Standards

Calibration and audit standards are of obvious and primary importance to quality assurance of pollution measurements. Development of appropriate and practical pollutant concentration standards of adequate accuracy, reliability, and availability has been a consistently high priority QAD project since the earliest needs for air pollutant measurements. Associated needs for calibration equipment, systems, devices, procedures, and verification techniques have also been addressed.

A fundamental need for primary gaseous pollutant concentration standards has been largely met by the availability of Standard Reference Materials (SRMs) from the National Bureau of Standards (NBS). Most EPA regulations require that pollutant concentration standards be traceable to such NBS primary standards. However, the cost and very limited supply of NBS SRMs makes them inappropriate for direct use by the many agencies and programs needing NBS-traceable standards. To alleviate this problem, NBS and EPA have jointly established a program to allow gas manufacturers to produce and sell high-quality gas concentration standards called Certified Reference Materials (CRMs). These CRMs are prepared and certified according to explicit CRM procedures developed by NBS and EPA and are considered equivalent to SRMs for purposes of establishing traceable working standards used for calibration and audits. Many of the EPA monitoring regulations have been amended to specifically allow working pollutant standards to be traceable to a CRM in lieu of an SRM.

The CRM certification procedure requires that a gas manufacturer first prepare a batch of ten or more standard cylinders, all having the same concentration, which must be within 1% of an appropriate SRM. From this batch, the manufacturer selects at least ten cylinders, which are tested for homogeneity. The batch of standards are then stored for at least two months (one month for CO standards), after which time all cylinders are reanalyzed to demonstrate stability and to determine the assay concentration for the batch. Two cylinders are selected at random for audit analysis by EPA. The results of all tests—including the EPA tests—are then sent to NBS to determine whether the entire batch of standards qualifies

---

[3] National Technical Information Service, Springfield, VA 22161.

TABLE 6—*List of Technical Assistance Documents available.*

Title	Author(s)	EPA No.	NTIS Stock No.	Approximate Cost
Technical Assistance Document for the Chemiluminescence Measurement of Nitrogen Dioxide	E. C. Ellis	EPA-600/4-75-003	PB268456	$9.50
Use of the Flame Photometric Detector Method for Measurement of Sulfur Dioxide in Ambient Air	W. C. Eaton	EPA/600/4-78-024	PB285171	$14.00
Technical Assistance Document for the Calibration of Ambient Ozone Monitors	R. J. Paur F. F. McElroy	EPA-600/4-79-057	PB80-149552	$8.00
Transfer Standards for Calibration of Ambient Air Monitoring Analyzers for Ozone	F. F. McElroy	EPA-600/4-79-056	PB80-146871	$14.00
Technical Assistance Document for the Calibration and Operation of Automated Ambient Non-Methane Organic Compound Analyzers	F. W. Sexton R. M. Michie, Jr. F. F. McElroy V. L. Thompson	EPA-600/4-81-015		
Investigation of Flow Rate Calibration Procedure Associated with the High Volume Method for Determination of Suspended Particulates	F. Smith P. S. Wohlschegel R. S. C. Rogers	EPA-600/4-78-047	PB291386	$14.00
Technical Assistance Document for Sampling and Analysis of Toxic Organic Compounds in Ambient Air	R. M. Riggin	EPA-600/4-83-027	PB83-239020	$16.95

TABLE 7—*EPA Standard Reference Photometer (SRP) network.*

SRP Site	SRP No.	Participating Agency	Installation Date
RTP, NC	1	U.S. EPA, EMSL/RTP	June 1984
Edison, NJ	3	U.S. EPA, region II	August 1984
Sacramento, CA	4	California Air Resources Board	September 1984
Chicago, IL	6	U.S. EPA, region V	March 1985
Houston, TX	5	U.S. EPA, region VI	May 1985
Denver, CO	8	U.S. EPA, region VIII	March 1986

to be sold as CRMs. If so, each standard cylinder sold from the batch is accompanied by joint NBS/EPA documentation listing the pertinent assay information and authenticating the cylinder as a CRM. A current list of CRMs is available from the Quality Assurance Division of EPA/EMSL.

As just noted, all gaseous pollutant standards used for calibration and auditing in state compliance networks are required to be traceable to either an NBS SRM or a CRM. The definition of "traceable" allows no more than one intermediate standard between the primary standard (SRM or CRM) and the field working standard.

The two most commonly used types of gaseous pollutant standards are cylinders of compressed gas and permeation devices. Procedures for certifying a working standard cylinder against a primary standard cylinder, and a working standard permeation device against either a primary standard cylinder or a primary standard permeation device are provided in Section 2.0.7 of the *Quality Assurance Handbook* (Volume II). Basically, these procedures give detailed guidance for setting up special certification systems designed to minimize comparative error, methods for precise calibration of analytical equipment, techniques for accurate comparisons of the two standards, appropriate tests for stability, and recommended recertification requirements. This section is currently under revision, and the updated version of these certification procedures should be available soon.

One criteria pollutant for which no SRM is available is ozone. Because of their instability, ozone concentration standards must be generated and certified locally as needed. In another joint EPA-NBS program, a highly stable, very precise, computer-controlled Standard Reference Photometer has been developed to assay ozone concentration standards and serve as a primary NBS reference standard for ozone. The EPA has established a network of these Standard Reference Photometers, currently consisting of the six locations listed in Table 7. Two additional locations are scheduled for installation in 1987. This network provides a means for state and local air monitoring agencies to compare their ozone standards to an authoritative primary ozone standard maintained and operated under closely controlled conditions. Such comparisons may be used to verify the accuracy of the agency's own local primary ozone photometer, verify certified transfer standards, or (in some cases) to certify transfer standards. The Network also provides EPA with information to assess the nationwide reliability and comparability of local ozone standards.

Two additional Standard Reference Photometers, one located at NBS and the other located at EPA/Research Triangle Park (EPA/RTP) are also maintained and directly intercompared once each year. The NBS instrument serves as a primary reference for NBS and is used to certify new Standard Reference Photometers. The EPA unit is used to recertify the Network instruments by direct intercomparison each year, using a recertification protocol developed jointly by EPA and NBS. This protocol requires (1) initial comparison of the EPA primary standard with the RTP Network instrument, (2) comparison of the EPA standard with each of the Network instruments at the Network site, and (3) final recom-

TABLE 8—*Verification of primary O_3 standards (photometers).*

Model	Number of Comparisons	Number of Units	Number of Comparison Results within Indicated Agreement			
			0 to 1%	1 to 2%	2 to 3%	>3%
Dasibi 1003	26	15	10	8	6	2
Dasibi 1008	22	19	7	6	2	7
CSI 3000	5	4	2	3	. . .	. . .
TECO 49PS	2	2	1	1	. . .	. . .
Totals	55	40	20	18	8	9[a]

[a] Five comparison results >5%.

parison of the EPA standard with the RTP Network instrument. Each recertification comparison requires simultaneous assays by each instrument of three sets of 12 ozone concentrations between 0 and 1000 ppb, with each concentration being assayed at least nine times. Target specifications for these recertification comparisons are 1% agreement for the regression slope and 3 ppb agreement for the intercept. The recertification data for each recertification are submitted to NBS for review, and if acceptable, an official NBS recertification document for the instrument is issued. All recertifications to date have met the target specifications.

The Standard Reference Photometer Network also provides data on the comparisons of state and local agencies' photometers and transfer standards to the Network Standard Reference Photometer. To date, results of these comparisons have been reported for 55 verifications of local primary standards and for 31 verifications of certified transfer standards. These results are listed in Tables 8 and 9. Forty-six of the local primary standards (84%) were within the established acceptable range of 3% agreement for primary standards, and all 31 of the transfer standards (100%) were within the 5% agreement established as acceptable for transfer standards.

TABLE 9—*Verification of O_3 transfer standards.*

Model (type)	Number of Comparisons	Number of Units	Number of Comparison Results within Indicated Agreement			
			0 to 1%	1 to 2%	2 to 3%	3 to 5%
Dasibi 1003 (photometer)	11	8	4	5	2	. . .
Dasibi 1003 (analyzer)	2	2	2	. . .	. . .	. . .
Dasibi 1008 (photometer)	12	6	3	3	5	1
TECO 49PS (photometer)	4	2	1	. . .	2	1
Monitor Labs 8810 (analyzer)	1	1	1	. . .	. . .	. . .
Monitor Labs 8500 (generator)	1	1	1	. . .	. . .	. . .
Totals	31	20	12	8	9	2

Performance Assessment

The third major area of quality assurance support by QAD for state and local monitoring agencies is assessment and evaluation of the quality of the monitoring data being routinely collected. With the establishment in 1979 of regulations prescribing uniform requirements for monitoring network design, siting, monitoring methods, quality assurance, and reporting methods [6], the quality of monitoring data reported could finally be documented in a uniform and meaningful way. These data quality assessments can now be used for much more reliable interpretation and evaluation of the air pollutant measurements being obtained.

Precision and Accuracy Reporting System

The data quality assessment provisions in Part 58 (Appendix A) [6] of the monitoring regulations specify explicit requirements for how each type of measurement system is to be assessed, how often, and how the results are to be reported. Data quality is defined primarily in terms of accuracy and precision. Accuracy is generally assessed (audited) by introducing NBS-traceable pollutant concentration standards into the measurement system and comparing the measured value to the true (standard) concentration. In some cases (such as for particulate matter), pollutant standards cannot be obtained for field use, and therefore only a portion of the measurement system (such as the flow metering system) can be reasonably audited.

Precision is assessed by comparing replicate measurements from the measurement system. For automated methods (analyzers), a fixed, low-level concentration standard is measured at least biweekly. For manual methods, simultaneous replicate measurements are obtained from duplicate (collocated) samplers operated periodically. The specific requirements and procedures used vary somewhat to accommodate the different pollutants and different types of measurement methods. Table 10 summarizes the current minimum data assessment requirements for SLAMS monitoring networks.

Precision and accuracy estimates are first calculated for each measurement system and then averaged over each calendar quarter. These estimates are then combined with the estimates from all of the similar measurement systems in each reporting organization to form composite estimates of precision and accuracy for the reporting organization. A reporting organization is defined as a state, subordinate organization within a state, or other organization that is responsible for a set of monitoring stations that monitor the same pollutant and for which precision or accuracy assessments can be appropriately pooled. The composite precision and accuracy estimates representing all the similar measurement systems in the reporting organization are expressed as 95% probability limits forming a probability interval. The upper and lower probability limits are determined as the mean value plus and minus 1.96 times the standard deviation.

Prior to 1986, states were required to calculate and report these composite or integrated precision and accuracy estimates for each of their reporting organizations. However, the regulations have been modified in 1986 to instead require the states to report the individual results from each precision or accuracy test carried out and from all collocated measurements obtained. The EPA/EMSL will then calculate the pooled average precision and accuracy 95% probability limits for each reporting organization. This change will allow EPA to have access, for the first time, to the individual precision and accuracy test results for each measurement system. This will facilitate detailed analysis of the data quality at specific sites or types of sites, for specific methods or types of methods, under specific conditions, etc. This additional information will greatly augment the description of specific blocks of the

TABLE 10—*Minimum data assessment requirements.*

Method	Assessment Method	Coverage	Minimum Frequency	Parameters Reported
PRECISION				
Automated methods for SO_2, NO_2, O_3, and CO	response check at concentration between 0.08 and 0.10 ppm (8 and 10 ppm for CO)	each analyzer	once per 2 weeks	actual concentration and measured (indicated) concentration
Manual methods including lead	collocated samplers	1 site for 1 to 5 sites 2 sites for 6 to 20 sites 3 sites for >20 sites (sites with highest concentration)	once per week	two concentration measurements
ACCURACY				
Automated methods for SO_2, NO_2, O_3, and CO	response check at 0.03 to 0.08 ppm[a] 0.15 to 0.20 ppm[a] 0.35 to 0.45 ppm[a] 0.80 to 0.90 ppm[a] (if applicable)	1. each analyzer 2. 25% of analyzers (at least 1)	1. once per year 2. each calendar quarter	actual concentration and measured (indicated) concentration for each level
Manual methods for SO_2 and NO_2	check of analytical procedure with audit standard solutions	analytical system	each day samples are analyzed, at least twice per quarter	actual concentration and measured (indicated) concentration for each audit solution
TSP	check of sampler flow rate	1. each sampler 2. 25% of samplers (at least 1)	1. once per year 2. each calendar quarter	actual flow rate and flow rate indicated by the sampler
Lead	1. check sample flow rate as for TSP 2. check analytical system with Pb audit strips	1. each sampler 2. analytical system	1. include with TSP 2. each quarter	1. same as for TSP 2. actual concentration and measured (indicated) concentration of audit samples (μg Pb/strip)

[a] Concentration $\times$ 100 for CO.

TABLE 11—*Percent completeness of precision and accuracy data for SLAMS.*

Pollutant	1981	1982	1983	1984
CO	77	89	91	89
SO$_2$	82	93	89	92
NO$_2$	56	72	84	86
O$_3$	83	89	91	90
TSP	94	97	98	97
Pb	. . .	. . .	81	82

SLAMS monitoring data and allow more detailed analysis and utilization of the data. The more detailed information will also be useful for monitoring the performance or adequacy of specific methods or types of monitoring instruments.

Precision and accuracy information for SLAMS monitoring networks is currently available for the years 1981 through 1984 [7–10]. The completeness of the data reported is shown in Table 11. Table 12 lists the percent of reporting organizations reporting probability intervals (absolute difference between the upper and lower 95% probability limits) less than 21% and less than 31% for years 1981, 1982, and 1983. In general, these tables show continued improvement in data quality for the SLAMS network.

National Performance Audit Program

Another data quality assessment facility established with the 1979 monitoring regulations in 40 CFR 58, Appendix A [6], was the National Performance Audit Program. This program provides assessments of SLAMS routine monitoring data accuracy similar to those obtained in the precision and accuracy reporting system, but with a greater degree of independence. Under this program, in which SLAMS monitoring agencies are required to participate, standards or test materials whose concentration or value is unknown to the analyst or to

TABLE 12—*Percent of SLAMS reporting organizations with 95% probability intervals[a] less than 21 and 31%.*

	1981		1982		1983	
	<21%	<31%	<21%	<31%	<21%	<31%
	PRECISION					
CO	78	92	82	96	87	98
SO$_2$	41	78	40	83	46	88
NO$_2$	36	70	40	71	40	85
O$_3$	59	87	62	91	62	93
TSP	45	73	43	74	50	80
Pb	. . .	. . .	46	59	39	64
	ACCURACY					
CO	75	88	86	94	84	94
SO$_2$	48	70	49	86	54	76
NO$_2$	50	76	49	70	56	76
O$_3$	63	87	72	87	75	91
TSP	86	97	78	95	90	97
Pb	. . .	. . .	87	100	89	98

[a] Probability interval = upper limit − lower limit.

TABLE 13—*National performance audit program annual schedule.*

Pollutant	Number of Audits
SO$_2$ (bubbler)	1
NO$_2$ (bubbler)	1
CO (monitor)	2
SO$_2$ (monitor)	continuous
NO$_2$ (monitor)	continuous
SO$_4$ (filters)	2
NO$_3$ (filters)	2
Pb (filters)	2
High volume (flow)	1
Acid rain	2

the participating agency are supplied by EPA. Measurements of the test materials are returned to QAD for determination of the performance results. The greater independence of this assessment provides an important check on the validity of the more detailed information from the precision and accuracy reporting system. Table 13 indicates the types of measurement systems covered by the National Performance Audit Program and the number of audits conducted each year. Results from this audit program are available for 1981 through 1984 [7–10].

Regional Systems Audits

A third key program in EPA's quality assurance support efforts is the program of systems audits of SLAMS monitoring programs. This program of systems audits is also authorized by the monitoring regulations in 40 CFR 58, Appendix A [6]. A systems audit is a complete and thorough inspection, review, and evaluation of a state or local agency's ambient air monitoring program to assess its compliance with established regulations governing the collection, analysis, validation, and reporting of ambient air quality data. A systems audit of each state or autonomous agency within an EPA Region is carried out annually by a member of the Regional Quality Assurance (QA) staff. Included in the systems audit are appraisals of such areas as monitoring network design and management, resources and facilities, field operations, laboratory operations, data management and reporting, quality control, and data quality assessment and reporting. This comprehensive review of the entire monitoring system helps to reveal problem areas that may limit or compromise data quality, and provides specific guidance and instruction directly to the agency for any corrective actions or improvements needed.

Summary

The EPA plays a key role in the quality assurance program for ambient air measurements in three major areas:

1. providing standardized and validated monitoring methods,
2. providing monitoring guidance and technical assistance, and
3. conducting assessments of the quality of the monitoring data.

While state and local air monitoring agencies carry out the monitoring and quality control programs necessary to collect the ambient data, effective team work between EPA and the air monitoring agencies ensures the high quality of those monitoring data.

References

[1] *Code of Federal Regulations*, Title 40, Part 50, "National Primary and Secondary Ambient Air Quality Standards," July 1985.

[2] *Code of Federal Regulations*, Title 40, Part 53, "Ambient Air Monitoring Reference and Equivalent Methods," July 1985.

[3] *List of Designated Reference and Equivalent Methods*, Quality Assurance Division, Environmental Monitoring Systems Laboratory, U.S. Environmental Protection Agency, Research Triangle Park, NC, April 1986.

[4] "Compendium of Methods for the Determination of Toxic Organic Compounds in Ambient Air," EPA-600/4-84-041, Quality Assurance Division, U.S. Environmental Protection Agency/Environmental Monitoring Systems Laboratory, Research Triangle Park, NC, April 1984.

[5] *Quality Assurance Handbook*, Vols. I–V, Quality Assurance Division, U.S. Environmental Protection Agency/Environmental Monitoring Systems Laboratory, Research Triangle Park, NC, 1976.

[6] *Code of Federal Regulations*, Title 40, Part 58, "Ambient Air Quality Surveillance," July 1985.

[7] Rhodes, R. C. and Evans, E. G., "Summary of Precision and Accuracy Assessments for State and Local Air Monitoring Networks, 1981," EPA 600/4-84-032, U.S. Environmental Protection Agency, Washington, DC, June 1983.

[8] Rhodes, R. C. and Evans, E. G., "Summary of Precision and Accuracy Assessments for State and Local Air Monitoring Networks, 1982," EPA 600/4-85-031, U.S. Environmental Protection Agency, Washington, DC, April 1985.

[9] Rhodes, R. C. and Evans, E. G., "Precision and Accuracy Assessments for State and Local Air Monitoring Networks, 1983," EPA 600/4-86/012, U.S. Environmental Protection Agency, Washington, DC, Feb. 1986.

[10] Rhodes, R. C. and Evans, E. G., "Precision and Accuracy Assessments for State and Local Air Monitoring Networks, 1984," EPA 600/4-86/031, U.S. Environmental Protection Agency, Washington, DC, Aug. 1986.

Robert B. Denyszyn[1] and Tom Sassaman[1]

Parts-Per-Billion Gaseous Mixtures: A New Challenge

REFERENCE: Denyszyn, R. B. and Sassaman, T., **"Parts-Per-Billion Gaseous Mixtures: A New Challenge,"** *Sampling and Calibration for Atmospheric Measurements, ASTM STP 957,* J. K. Taylor, Ed., American Society for Testing and Materials, Philadelphia, 1987, pp. 101–109.

ABSTRACT: Specialty gas manufacturers are being driven by regulatory requirements to manufacture low concentration gaseous mixtures of various air pollutants. This paper describes some of the basic difficulties incurred in preparing parts-per-billion (ppb) concentrations of various volatile organics such as pyridine, Freon®113, Freon®114, vinylidene chloride, toluene, 1,4 dioxane, acetone, chlorobenzene, acetonitrile, and methylethyl ketone. Concentrations range from 10 to 150 ppb. During the manufacturing and certification of these mixtures, several problems were encountered, including:

1. Ghost peaks due to gas chromatographic system background.
2. Adsorption of compounds in the analytical system as well as on cylinder walls.
3. Difficulties in preparation of primary standards for calibration.
4. Contamination of cylinders, even after special cleaning and preparation.

It is therefore imperative that when purchasing such mixtures, a certificate of analysis as well as a copy of the chromatographs be provided to the user.

KEY WORDS: toxic organic volatile, parts-per-billion mixture, stability, adsorption, air pollutants, sampling, air quality, calibration, atmospheric measurements

Specialty gas manufacturers are being driven by regulatory requirements to manufacture very low concentration gaseous mixtures of various air pollutants. This paper describes some of the basic difficulties incurred in preparing parts-per-billion (ppb) concentrations of various volatile organics such as pyridine, Freon®113, Freon®114, vinylidene chloride, toluene, 1,4-dioxane acetone, chlorobenzene, acetonitrile, and methylethyl ketone. Concentrations range from 10 to 150 ppb. The problems encountered in preparation of these standards include:

1. Ghost peaks due to gas chromatographic system background.
2. Adsorption of compounds in the analytical system. In some cases, the problem is compounded by adsorption of the compound by the cylinder wall.
3. Difficulties in the preparation of primary standards for calibration.
4. Contamination of cylinders, even after special cleaning and preparation.

For many years, the need for gaseous mixtures was limited to combustion-related by-products such as carbon monoxide (CO), carbon dioxide (CO_2), oxides of nitrogen (NO_x), sulfur dioxide (SO_2), and hydrocarbons. The hydrocarbons were, in most cases, limited to

---

[1] Manager, Research and Development, and laboratory technician, respectively, Scott Specialty Gases, Plumsteadville, PA 18949.

propane and methane. On a few occasions, the need for other gas mixtures occurred, primarily in the field of industrial hygiene.

In the late 1970s, several ambient air monitoring programs near industrial centers showed the presence of organic volatiles ranging from the typical automotive exhaust hydrocarbons to the more industrially utilized chemicals such as perchloroethylene, carbon tetrachloride, etc. [1–5]. It was not until 23 January 1981 that the U.S. Environmental Protection Agency (EPA), under the Resource Conservation and Recovery Act (RCRA), promulgated the statutory required standards for hazardous waste incinerators. The main thrust of the legislation required that hazardous materials be 99.99% destroyed or removed by the process. In order to meet such requirements, incinerators must perform a trial burn evaluation. During the evaluation, the waste must be reduced to levels as low as low ppb concentrations in the gaseous effluent. Knowing the outlet concentration of the incinerator is crucial, because this establishes if the process is performing in accordance with the statutory requirements. These very low concentrations presented a new challenge to the analytical chemist who was now being asked to analyze very low concentrations relatively quantitatively.

To evaluate the sampling method, as well as the analytical equipment, accurate standard gas mixtures at the ppb level are required. Parts-per-billion level gaseous products were not new to Scott Specialty Gases. Scott produced gaseous mixtures of SO_2, NO_x, CO_2, CO, and many other gases such as propane, and methane. The hazardous organics presented a real challenge since a great many more problems can occur with organic volatiles than with inorganic gases.

During the past several years, Scott Specialty Gases has developed quality control and manufacturing practices that enable its manufacturing facilities to prepare either simple or complex organic volatile mixtures at the ppb level. This process is called Micrograv®.

Procedure

For many years, the specialty gas industry limited the certification of organic volatile concentrations to 1 ppm. The major reason for this limit was due to the system typically used to quantitatively analyze such concentrations. The popular flame ionization detector (FID) can usually perform a 1-ppm analysis with a precision of ±1 to 2%. The FID loses sensitivity with halogenated organics, and a concentration usually must be between 5 to 10 ppm in order to achieve the same precision.

Halogenated organics can be analyzed using the electron capture detector (ECD) down to the parts-per-trillion (ppt) level, but the response of most instruments becomes significantly nonlinear above the 10 to 100 ppb level. Other detectors are available such as the Hall detector and the photoionization detector, but their long-term reliability can vary. Therefore, Scott Specialty Gases retained the concentration mechanism and the FID as its primary method for analyzing organic volatiles, with the ECD as a secondary method for very low concentrations (<50 ppb), or weakly responding species such as mono and di substituted halocarbons.

The cryogenic concentration technique utilized to enhance the detector signal is a well developed technique. Some of the major aspects of the technique include the following.

1. The method is virtually universal—it concentrates all organic volatiles greater than methane.
2. The collection efficiency is very high, not only because of the fundamental principles, but also because the matrix is very clean—no water, no CO_2, no reactive species such as O_3 and NO_x.

3. The recovery is very high for most compounds exceeding 95%.
4. It is cost effective.

The cryogenic technique used in this work is similar to the one used by Denyszyn et al. [6] to measure trace hydrocarbons downwind of urban centers, and similar to the system used by Lonneman [7] and Cox [8]. The technique has been improved since the mid-1970 design. Liquid oxygen has been replaced with liquid argon. This cryogenic fluid provides a much higher degree of safety than liquid oxygen. Changing from liquid oxygen to liquid argon has not affected the collection efficiency. Pressure measurements are made using electronic pressure transducers providing greater precision for the volume collection. With a sample volume of 300 mL, the volume uncertainty is ±0.02%. The materials in contact with the sample have been restricted to nickel and 316 stainless steel. All metal parts are both silanized and heated to 175°C to eliminate adsorption of the minor components. The organic volatiles are released from the trap by heating it at least 20 to 30°C above the highest boiling point of any minor component in the mixture. All components were separated as a 1.8 m by 3.175 mm (6 ft by ⅛ in.) stainless steel column packed with ⁸⁰/₁₀₀ mesh Carbopack C coated with 0.1% of SP-1000. The column was temperature programmed from 35 to 190°C at 10°C/min with an initial hold of 6 min. Upon reaching 190°C, the temperature was held for 2 min.

Results and Discussion

Recent work with the following organics: vinylidene chloride, Freon 113, Freon 114, pyridine, acetone, dioxane, toluene, and chlorobenzene showed significant problems such as:

1. Extraneous peaks.
2. Gas chromatographic system contaminations.
3. Adsorption of minor components to both column and cylinders.
4. Contamination of the high pressure cylinders.

These difficulties could, if not properly addressed, degrade the value of the standard purchased by the customer. These difficulties however can be minimized through careful analytical work.

The initial chromatograms of the eight-component gaseous mixture showed ten major peaks and several minor peaks (Fig. 1). The extra peaks were definitely a significant problem

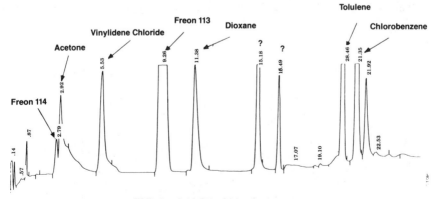

FIG. 1—*Initial analysis of mixture.*

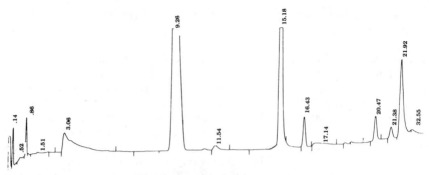

FIG. 2—*Initial nitrogen (N₂) blank.*

since theoretically, only eight components were introduced. The peak at 9.26 min was significantly too large in comparison to other peaks, and the peak at 2.92 min had too much tailing to quantify.

The first major problem to resolve is the total number of peaks in the chromatogram. One common problem encountered in ppb analyses is system contamination. To evaluate this potential, several chromatograms were run with a cylinder containing ultra-high-purity nitrogen. One of the first chromatograms is shown (Fig. 2). The chromatogram showed a significant peak at 9.26, 15.16, and 21.92 min with minor peaks at 16.49 and 20.46 min. The 9.26-min peak happened to be the retention time for Freon 113. The peak at 15.16 min was attributed to the silanizing agent used on the inlet tubing. The system background was reevaluated after increasing the temperature on both the sampling valve and inlet tubing, plus purging the system with ultra-high-purity nitrogen (Fig. 3). The Freon 113 peak at 9.26 min was reduced to 500 ppt, the peak at 16.49 min was eliminated, and while the peak at 20.47 min was never eliminated, it was reduced to the equivalent of approximately 1 to 2 ppb.

The analysis cycle consists of two steps: the concentration step and the chromatographic step. During the concentration step, the column oven temperature is reduced to the initial temperature (that is, 35°C), and the carrier gas continues to flow through the column and detector. Should any impurities be present in the carrier gas, these impurities could concentrate on the column and, when the oven temperature program is initiated, these materials could be released and detected as part of the sample. Therefore, it is important to measure the amount of organic generated by the chromatographic system. In our case, a small peak at 20.40 min was detected and found to be reproducible (Fig. 4). This peak coincided with

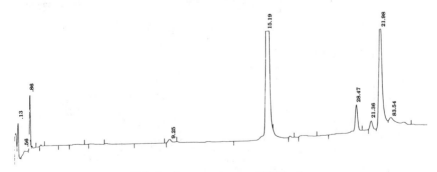

FIG. 3—*Improve nitrogen (N₂) blank.*

FIG. 4—*Chromatographic system blank.*

the elution of toluene. Had this background analysis not been performed, the results would have a bias.

Figure 1 showed acetone eluting at 2.92 min. The geometry of the peak was such that quantification would be impossible. Therefore, a new column was selected; a column that was not as sensitive to adsorption. A glass column with the same packing was evaluated and superior performance was obtained (Fig. 5). The chromatogram showed only seven peaks; pyridine was missing. It was initially thought that the peak at 16.49 min was pyridine. Work at the ppm level, with a stainless steel column, showed a retention time of approximately 17 min for pyridine. On the glass column, relative retention time calculation indicated a theoretical peak at 18.51 min for pyridine. The chromatogram of Fig. 5 showed no peak at that retention time, indicating the possibility that either the pyridine was not behaving ideally in the cylinder, or the analytical system did not allow for the retrieval of that molecule from the cylinder. The nonideal behavior of pyridine at the ppb level is somewhat surprising since our experience at the ppm level did not indicate such a problem.

To confirm the concentration of standards in high pressure cylinders, high quality primary standards must be generated. Several methods may be used, three of which are discussed here.

Method 1—Prepare (usually gravimetrically) gaseous standards in similar containers at the same or slightly varying concentrations. This practice is satisfactory for gaseous mixtures in the ppm range, but not at the ppb and ppt concentration levels. The main reason is the surface-to-gas interactions that are more apparent at these levels. Failure to recognize these effects could result in a significant bias in the concentrations associated with the standards. Surface-to-gas interactions can be minimized through surface treatment techniques, but it may not be assumed that such treatments are effective for all compounds.

Method 2—Use permeation devices to generate the required concentrations. This method has been used by many to generate primary standards. Permeation devices are utilized in a dynamic fashion and adsorption of the permeated component can usually be overcome through careful selection of the component in contact with the gas, but one must recognize the limitations.

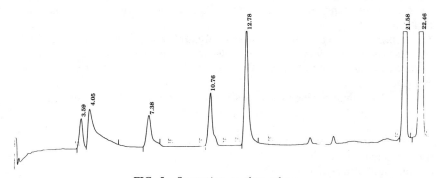

FIG. 5—*Separation on glass column.*

TABLE 1—*Surface-to-gas interactors.*

Material of Construction for Containers	Interaction
Chromium molybdenum steel Spun stainless steel Teflon coated steel Aluminum Acid washed glass Treated aluminum (Aculife™) Electropolished stainless steel Treated electropolished stainless steel (Aculife™) Silanized glass	MOST ↑ ⏐ ⏐ ↓ LEAST

(*a*) The resulting concentration generated by a permeation device is dependent on several variables: the permeation rate, the temperature of the permeation device, and the dilution flow rates.

(*b*) Permeation devices may require several months of stabilization prior to achieving a stable permeation rate. This makes it difficult to use in a production situation.

(*c*) Not all chemicals behave ideally in permeation devices, and permeation rate shifts are not unusual.

(*d*) The number of permeation rate devices required to service the need of the marketplace could be extremely large, possibly up to several thousand, if several devices are required to provide varying permeation rates.

Method 3—Static dilution of the minor component in "ideal" containers. At low concentrations, ppm and below, surface-to-gas interaction is one of the variables that one must always consider as being one of the uncertainties in any standard. Since there are a large variety of materials from which containers can be constructed, Table 1 shows the most to least likely materials to cause surface-to-gas interactions.

The use of silanized glass has been well demonstrated in chromatography where glass column adsorption could be minimized through the use of a silanizing agent. Glass containers (approximately 10 L) are used to prepare primary standard gaseous mixtures. These gas mixtures are prepared gravimetrically (that is, both minor and balance gases are weighed). At low concentrations (10 ppm), the minor components are dissolved in a suitable solvent. Selection of the solvent is crucial both in terms of chemical purity and chromatographic behavior. Several standards are prepared to bracket the concentration of the minor components of interest.

Not all problems are analytical. A significant number of problems can occur in the manufacturing process. Ideally, the manufacturing process should be accurate and precise enough to provide a product for which the analytical technique serves as a check on the manufacturing technique.

Some of the manufacturing problems associated with ppb level gaseous mixtures are:

(*a*) contamination,
(*b*) gas-to-surface interaction,
(*c*) mathematical errors,
(*d*) omission of a component, and
(*e*) wrong components.

TABLE 2—*Residual gases.*

Residual Gas Initial Concentration	Residual Gas Total Pressure, Pa	Final Concentration
100%	100	10 ppm
100%	10	1 ppm
100%	1.0[a]	100 ppb
1.0%	10	10 ppb
1000 ppm	10	1 ppb
1%	1.0	100 ppt[b]
1000 ppm	1.0	10 ppt

[a] Maximum vacuum attainable in cylinder.
[b] Must achieve to obtain <500 ppt.

Contamination is one of the most serious problems in preparing ppb level gaseous mixtures. The lower the concentration, the more significant the problem. If a mixture should be 10 ppb, a significant contaminant is a component that is 5% of the component concentration, or 500 ppt. A standard practice in specialty gases is to evacuate the cylinder to approximately 10 Pa. Under this condition, what residual impurities can be expected in a filled cylinder? Table 2 summarizes the expected concentrations.

The final concentration shown in Table 2 of the component assumes the cylinder is pressurized to 14 000 kPa. The table assumes that the residual gases are well behaved. It is certain that gas cylinders initially must be free of contaminant to ensure a product with 500 ppt contamination. Adsorbed gases can be removed via several mechanisms such as heating of the cylinder surface, displacement, or chemical reaction. The Scott Specialty Gases Aculife™ process uses three techniques.

Contamination can occur at all phases of the manufacturing cycle. For example, after a cylinder has been evacuated to <10 Pa, it may be contaminated with room air. This is especially so if the cylinder remained under vacuum for any period of time or room air is accidentally introduced into the cylinder. One must remember that workplace environments can contain several ppm of various halogenated species (that is, Freons, degreasing agents, solvents). Another possible source of contamination is the outgasing of polymeric components that may have been exposed to degreasing agents. Cylinder valves are especially prone to such problems. To provide contaminant-free products, dedicated blending systems must be used. Even under careful conditions, contamination can occur. Figure 6 is a chromatogram

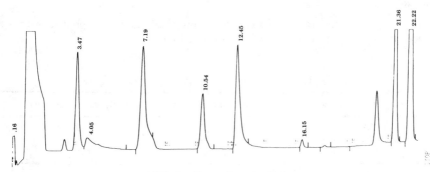

FIG. 6—*Contaminated cylinder.*

TABLE 3—*Adsorption of acetonitrile and methylethyl ketone.*

	Acetonitrile				Methylethyl Ketone			
	Aculife® Aluminum Treated Cylinders		Untreated Aluminum Cylinders		Aculife® Aluminum Treated Cylinders		Untreated Aluminum Cylinders	
Theoretical concentration, ppb	161	161	167	169	164	163	169	168
Analytical concentration, ppb	129	131	109	112	168	169	143	136
Analytical precision %	3.0		3.6		2.4		1.7	
%[a]	−14	−19	−35	−34	+2.5	+3.7	−16	−19
Theoretical concentration, ppb	15.4	15.1	15.5	15.9	15.5	15.8	16.0	16.3
Analytical concentration, ppb	14.0	13.0	8.9	8.3	15.8	16.3	7.68	7.06
Analytical precision, %	3.0		3.6		2.4		1.7	
%[a]	−9	−14	−43	−48	+2.0	+2.8	−53	−57

[a] Defined as $\dfrac{[\text{theoretical concentration}] - [\text{analytical concentration}]}{[\text{theoretical concentration}]} \times 100.$

from a contaminated cylinder. The contaminant concentration was approximately 200 ppb.

Gas-to-surface interactions are the most difficult to identify, and only through thorough testing and literature review can one establish such interactions. Some of these interactions may be weak, some may be strong. All are deleterions toward having a well-behaved standard.

A well-behaved gaseous standard is one that is not affected by time, temperature, and total pressure of the mixture.

Adsorption mechanisms were observed with some chemical systems. Mixtures of acetonitrile and methylethyl ketone were prepared. The concentrations desired by the customer were 150 and 15 ppb. The mixtures were prepared in two types of aluminum cylinders: Aculife® treated and untreated. Two cylinders were prepared at each concentration level. Each mixture was analyzed twice, with a month elapsing between each analysis. The precision of the measurement is shown in Table 3.

The results are fairly conclusive. Cylinders that are not treated show significantly lower concentrations of organic volatiles. In the case of the acetonitrile, the gravimetric values of the treated cylinders are in agreement with the analytical measurement with virtually the same uncertainty. The untreated cylinders show significantly lower concentrations.

Conclusion

Gas mixtures of many organic volatiles can be prepared at the low ppb level with blend specifications and analytical uncertainties shown in Table 4.

TABLE 4—*Parts-per-billion mixture specifications.*

Concentration Range, ppb	Blend Specifications	Analytical Uncertainty[a]
999 to 500	±2%	±2%
499 to 100	±5%	±5%
99 to 10	±10%	±10%

[a] At the 95% confidence level.

Unlike ppm mixtures of similar chemicals, a variety of problems can be encountered such as:

1. Gas chromatographic background.
2. Adsorption of compound.
3. Preparation of primary standards.
4. Contamination of cylinders.

Most of these problems can be overcome with stringent manufacturing practices and sound analytical work. It is certain that no ppb mixture should be ordered without a confirmatory analysis. It is also recommended that customers requesting such mixtures should be provided with a copy of the confirmatory chromatograms.

References

[1] Pellizzari, E. D., "Analysis of Organic Air Pollutants by Gas Chromatography and Mass Spectrometry," U. S. Environmental Protection Agency, EPA-600/277-100 and EPA-600/2-79-057, Washington, DC, June 1977.
[2] Pellizzari, E. D. and Bunch, J. E., "Ambient Air Carcinogenic Vapors: Improved Sampling and Analysis Techniques and Field Studies," U. S. Environmental Protection Agency, EPA-600/2-79-081, Washington, DC, May 1979.
[3] Pellizzari, E. D., Bunch, J. E., Berkley, R. E., and McRae, J., *Analytical Letters,* Vol. 9, 1976, p. 45.
[4] Pellizzari, E. D., "Measurement of Carcinogenic Vapors in Ambient Atmospheres," U. S. Environmental Protection Agency, EPA-600/7-78-061, Washington, DC, April 1978.
[5] Pellizzari, E. D., Erickson, M. D., and Zweidinger, R. A., "Formulation of a Preliminary Assessment of Halogenated Organic Compounds in Man and Environmental Media," U. S. Environmental Protection Agency, EPA-560/1379-066, Washington, DC, July 1979.
[6] Denyszyn, R. B., Plessor, D., and Cooley, C. C., "Improving the Reliability of Low PPM Carbon Monoxide Gaseous Standards," presented at the 34th Pittsburgh Conference on Analytical and Applied Spectroscopy, Atlantic City, 8 Mar. 1983, Paper 331.
[7] Lonneman, W., Kopczynski, S., Darley, P., and Sutterfield, F., *Environmental Science and Technology,* Vol. 8, 1974, p. 229.
[8] Cox, R. D. and Earp, R. F., *Analytical Chemistry,* Vol. 54, 1982, pp. 2265–2270.

Gerald D. Mitchell[1]

Trace Gas Calibration Systems Using Permeation Devices

REFERENCE: Mitchell, G. D., **"Trace Gas Calibration Systems Using Permeation Devices,"** *Sampling and Calibration for Atmospheric Measurements, ASTM STP 957*, J. K. Taylor, Ed., American Society for Testing and Materials, Philadelphia, 1987, pp. 110–120.

ABSTRACT: Permeation devices are used in a variety of dynamic gas dilution systems to prepare trace or ultra-trace mixtures of gases and vapors, generally for calibration purposes. The sources of errors in the use of a permeation device for calibration are associated with the reproducibility of flow measurements, temperature control, and the stability of the mass transport characteristics of the permeation device. Since the primary component of such a single or multi-stage trace gas calibration system is the permeation device, then it is befitting that a comprehensive study of the stability and transport properties be undertaken. This paper describes trace calibration systems and the factors that influence their operation. Also presented is COGAS, a dynamic dilution system that serves to calibrate both permeation devices and instruments. Such a system improves precision, reduces manpower requirements, and is versatile in its application.

KEY WORDS: permeation, gas diffusion, calibration, solubility, dynamic dilution, gas analysis, sulfur dioxide, sampling, air quality, atmospheric measurements

Sources of accurate, stable low-level gas calibration mixtures are extremely important to the environmental scientist. The development and testing of analytical methods for determining gaseous air pollutants and trace levels of specific components are expedited by the availability of accurately known gas calibration mixtures. In addition, atmospheric and environmental measurements require that the test method and instrumentation yield reliable and repeatable results when carried out by different laboratories. The uncertainty of the results generally can be reduced when laboratories calibrate their method and instrumentation with the same reference mixture and procedure such as, ASTM Recommended Practice for Calibration Techniques Using Permeation Tubes (D 3609-79).

Today, there are primarily two techniques for the preparation of calibration gas mixtures: (1) those that are statistically prepared and stored in cylinders, and (2) those that are dynamically prepared by using permeation or diffusion devices. The advantages of using a dynamic system, such as those involving permeation devices, are that the adverse phenomena of static systems are avoided and that calibration systems based on permeation devices are especially useful when working with reactive gas mixtures. If the component of interest is unstable or reactive, the products are swept away and are continually replaced by new and unreacted components. In addition, dynamic systems have a much wider range of applicability than fixed concentration standard mixtures.

The primary focus of this paper is on calibration systems and the physical precautions that must be considered when operating a trace calibration system with a permeation device.

---

[1] Research chemist, Gas and Particulate Science Division, Center for Analytical Chemistry, National Bureau of Standards, Gaithersburg, MD 20899.

TABLE 1—*Dynamic methods for producing controlled concentration gas mixtures.*

Method	Concentration Range	Accuracy, %
Single dilution[a]	100 ppm to 5.0%	1 to 3
Multi-dilution[a]	1 ppb to 1%	5 to 10
Diffusion	0.1 ppm to 500 ppm	2
Permeation[a]	0.2 ppm to 200 ppm	1 to 3
Electrolytic	low ppm to 1%	2 to 5
Chemical reaction[a]	low ppm to 1%	3 to 10

[a] Calibration systems in which permeation devices may be used.

In addition, a specific calibration system, the Computer Operated Gas Analysis System (COGAS), is described that may be used as both an instrument calibration system and a permeation device calibration system.[2]

Calibration Systems

Many of the gases currently of interest from an environmental perspective, that is, sulfur dioxide (SO_2), nitric oxide (NO), and nitrogen dioxide (NO_2), are reactive or toxic or both. Because of the reactivity and toxicity of such gases, it is advantageous to use calibration systems and methods that allow the dynamic production of the desired gas calibration mixture. Table 1 shows some of the dynamic systems that are in use today and the expected levels of accuracy. Of these systems, four may be used with permeation devices: namely, dynamic dilution (single and multi-stage), permeation, and chemical reaction. There are experimental difficulties for each. Emphasis here will be placed on the physical precautions that must be considered when setting up dynamic systems, particularly those with permeation devices.

Electrolytic Systems

Electrochemical methods have long been used in analytical chemistry (for example, electrodepositing metal ions from solution). Using such methods, many gases can be synthesized on a laboratory scale as a byproduct or side reaction of electrolytic or coulometric methods. Hersch [1] has demonstrated the utility of such methods for a variety of systems. However, when using the electrolytic method, the selection of an appropriate electrolyte involves several considerations. First, there should be no side reactions occurring at the electrode of interest. If gases at the second electrode interfere, then the design of the cell must allow for the diversion of these gases away from the carrier-gas stream. The time to attain the equilibrium concentration of the calibration mixture also must be minimized. A further consideration is that the rate of production of the desired gas mixture should be as linear as possible in order to produce a uniform and homogeneous calibration gas mixture. Because of its capability of producing small quantities of gas, this method is ideal for preparing calibration mixtures where ultra-low concentrations are required. However, these systems do not utilize a permeation device.

Permeation/Diffusion Systems

The use of permeation devices for the generation of calibration gas mixtures is widely practiced [2,3]. In a permeation device, the contained substance has the property of diffusing

[2] E. E. Hughes and J. E. Suddueth, private communication.

or permeating through the walls of a tube that is constructed of an organic polymeric material. A uniform rate of permeation is established if the temperature, concentration gradient, and the tube geometry remain constant. This phenomenon leads to a very versatile method of producing ultra-low concentrations of calibration gas mixtures. In preparing the device, the liquid form of the gas is sealed inside a polymeric container, typically, polytetrafluoroethylene (Teflon®). The test substance dissolves in the polymer and permeates through the wall of the tube at a rate that depends upon the temperature and the concentration gradient across the tube material. The substances that work best with this method are those that have a critical temperature above 20 to 25°C [4]. Typically, these include such materials as SO_2, NO_2, chlorine (Cl_2), hydrogen chloride gas (HCl), benzene, carbon disulfide, Freon-11, and numerous others.

The prime moving forces behind these methods (diffusion/permeation) are described by Fick's first law of diffusion, Henry's law of solubility, and the permeability, which relates the diffusion coefficient (D) to the solubility constant ($\overline{S}$) through the permeability coefficient (P). The relationship of diffusion solubility and permeation may be expressed as

$$P = \overline{S}D \qquad (1)$$

Permeation devices have several desirable advantages over other means of providing input to dynamic systems. With these devices, there is a low degree of complexity, the gaseous concentration that can be generated ranges from 0.1 to 200 ppm, they are ideal for field use, and high precision and accuracy may be obtained (±1%) [2].

Dynamic Dilution Systems

A dynamic gas dilution system has a constant, stable flow rate of diluted gas through it (which can be predictably varied), and also has a means of adding a controlled amount of a gas or vapor of interest. A variety of systems have been devised for injecting known, controlled amounts of the component of interest into the flowing diluent stream. At the low end of the concentration range (see Table 1) where multistage systems are used, consideration must be given to problems that may occur from the following:

1. The need for high quantities of diluent.
2. Flow rate control.
3. Back diffusion.
4. Surface absorption.
5. Gas blending.

In addition, multistage dilution techniques that require more than two dilution steps become considerably more difficult. As more dilution steps are added, the system pressure builds up, and the quantity of the gas of interest that is needed becomes the major concern. In such arrangements, as much as 99% of the test gas could be bled off and not used in the output stream.

Chemical Reaction Systems

Controlled addition of a gas of interest to the moving stream of a diluent also may be accomplished by a chemical reaction system. If the material of interest is inordinately reactive, unavailable commercially, or toxic, it may be desirable to produce it continuously as needed. Typical types of reactions that are used for this method of producing calibration mixtures of desired components are recombination, catalytic, thermal decomposition, rearrangements, and photochemical dissociation. The reactions may be carried out in either the gas or liquid phase. One such technique that is in extensive use in atmospheric chemistry is the gas-phase titration. This technique is uniquely suited for certain applications because of its ability to control the products of interest [5,6].

Physical Precautions

Regardless of which calibration system is chosen for a specific analytical problem, there are certain precautions that must be considered in order to assure accurate and reliable results. The precautions outlined here are not intended to be all-inclusive, but are presented in order to raise an awareness of their importance in trace calibration systems.

Homogeneity

Regardless of the precision and accuracy of the analytical results, the procedure used cannot be any better than the quality of the sample used. That is, the sample must be a representative blend of what is to be measured by the instrument. For each of the calibration systems discussed, the primary concern of the investigator must be directed at the collection of a representative sample of the calibration gas mixture. This concern causes the investigator to take certain steps in the design of the apparatus, and to be vigilant of possible sources of contaminants. Turbulent gas flow and mixing chambers are only two of many techniques that can be used to assure uniform calibration mixtures.

Great care is required to establish a uniform calibration mixture regardless of the kind of calibration system used. Effects that have a direct bearing upon a system's ability to produce a homogeneous mixture include absorption on walls, diffusion, sampling, and temperature.

Absorption

Losses in calibration mixtures may occur either by primary absorption on container walls or connecting tubing, or by secondary absorption that occurs through chemical reaction of the component of interest with previously absorbed material. This problem is most critical when considering trace levels of reactive gases such as NO_2 and SO_2 [7]. In dynamic systems, a prolonged equilibration time is required before a stable calibration mixture can be assured because of the potential problem of absorption.

Diffusion

Diffusion of the sample outward or a contaminant or even a major constituent inward is probably one of the most frequent sources of error in gas analysis. Losses due to diffusion are in addition to those due to absorption. The permeability of calibration systems to both the gases of interest and unwanted contaminants have to be determined if reliable results are to be obtained.

Mechanics of Sampling

The mechanical features of the sampling system can also introduce various errors if they go undetected. Sources of these errors are leaks, faulty seals, corrosion of vital parts, and the change in the calibration of sensors that may be directly attributed to the mechanics of the system.

Temperature

When evaluating a calibration system, temperature effects must be clearly delineated for the reference source being sampled (for example, permeation device) and for the measuring system itself. Extreme caution in the use of data must be exercised when a thermal change is determined to have a measurable effect upon the mechanical performance of a calibration system. For calibration systems that utilize permeation devices, there is a measurable change in the permeation rate as a function of temperature (that is, about 1% per 0.1°C).

Volumetric Errors

Since rate meters of all types require frequent calibration, consideration should be given to the portability of the systems, pressure drops across the measuring instrument, the temperature coefficient of the measuring instrument, and the reliability of the pumping system. For dynamic dilution systems, the critical element is the flow control system. Flow controllers must be independently characterized as to temperature, pressure, and other parameters they are subject to influence.

Observational Errors

Observational errors in the operation of a dynamic calibration system lie mainly in the inaccuracy or imprecision of the reading that is affected by mechanical movement, as in the case of rotameters, floats, and analog meters.

Diffusion, Permeation, and Solubility

If we now consider diffusion/permeation only in terms of calibration systems that utilize permeation devices (Table 1), one should consider not only the diffusion and permeability effects of the system, but also the parameters that affect the device itself.

From Fick's first law, the rate of diffusion, J, is given by the expression

$$J = -D \frac{dc}{dx} \tag{2}$$

where D is the diffusion coefficient, c is the concentration of the diffusing species, and x is the distance diffused. A variation in temperature affects the diffusion coefficient through a Hess' law relationship

$$\frac{d(\ln D)}{d(1/T)} = -E_D/R \tag{3}$$

where E_D is the diffusion activation energy, R is the ideal gas constant, and T is the temperature, °K. Under steady-state conditions in which the flux, J, is invariant with time, the integral value of J

$$\int_0^L J \, DL = \int_{C_1}^{C_2} D \, dC \tag{4}$$

$$J = -\frac{1}{L} \int_{C_1}^{C_2} D \ dC \tag{5}$$

is inversely proportional to the thickness (L) of the solid material. Equation 4 can be integrated when the variation of D with concentration is known. However, in the case where Van der Waals' forces predominate, both among the diffusing molecules and between the diffusing and substrate molecules, D varies with concentration. This condition usually exists when the solubility of the diffusing species in the substrate is high. Under such conditions, the variation of both the diffusion coefficient (D) and the diffusion activation energy (E_D) with concentration may be expressed by the following equations

$$D = Ae^aC \tag{6}$$

$$E_D = B - bC \tag{7}$$

where A and B are constants of the substrate and a and b are constants of the diffusing molecules.

When the intermolecular forces are weak and solubility is low, concentration variations do not have an effect upon the diffusion coefficient. Under these conditions the integrated form of Eq 4 becomes

$$J = -\frac{D}{L}(C_2 - C_1) \tag{8}$$

where C_1 and C_2 are concentrations of the dissolved, diffusing molecules at the two surfaces of the substrate. In the gas-polymer system, a Henry's law type equilibrium is exhibited between the external gas pressure and the concentration of the gas dissolved in the substrate. This is expressed as follows

$$C = \overline{S}P \tag{9}$$

where C is concentration, P is pressure, and $\overline{S}$ is the solubility coefficient for a particular gas-polymer system [8]. The coefficient ($\overline{S}$) is the reciprocal of Henry's law constant that is a special case of the Nernst distribution law. In the proper units, $\overline{S}$ is identical to the Bunsen adsorption coefficient [9] defined as the volume of gas in cubic centimeters, reduced to standard conditions, which dissolves in 1 cm^3 of substrate at a given temperature when the pressure is 1 atm. With Eq 9, it is assumed that (a) the polymer substrate behaves as a liquid or is amorphous in character; (b) conditions are isothermal; and (c) the system is in a state of local thermodynamic equilibrium. The variation of $\overline{S}$ with temperature is expressed by the Van't Hoff equation

$$\frac{d \ln \overline{S}}{dT} = \frac{\Delta H_s}{RT^2} \tag{10}$$

or

$$\frac{d \ln \overline{S}}{d(1/T)} = \frac{\Delta H_s}{R} \tag{11}$$

where ΔH_s is the enthalpy of solution of the dissolved molecules.

Whenever the surfaces of the polymer substrate are in equilibrium with the external gas pressure of the diffusing molecules, Eq 9 may be substituted into Eq 8 giving

$$J = -\frac{D\bar{S}}{L}(P_2 - P_1) \tag{12}$$

If Henry's law applies to the solubility of the permeator in the substrate and Fick's first law holds for the diffusion processes, then Eq 12 implies that the permeation rate is proportional to the differential partial pressure across the substrate and inversely proportional to the thickness of the substrate. In the case of permeation devices, Eq 12 may be expressed by

$$J = \bar{P}/L \, (P_2 - P_1) \tag{13}$$

where $\bar{P}$ is the permeability coefficient and is defined as $D\bar{S}$. This relationship

$$\bar{P} \equiv D\bar{S} \tag{14}$$

is the fundamental model for gas permeation in polymer substrates [10–12]. The relationship is considerably more complex than is initially considered. It relates the permeability coefficient ($\bar{P}$) to the phase-boundary equilibrium constant ($\bar{S}$) (a thermodynamic property), and to the mass-transfer coefficient (D) (a kinetic property).

Since permeation devices are at least binary systems consisting of a stationary polymer substrate and a gaseous or liquid component, and the permeation process depends on both kinetic and thermodynamic properties of the system, it is not unexpected to find that gas permeation through polymeric materials is not solely dependent upon a specific property of either the permeator or the substrate, but, rather, upon the interaction between them. Any theory developed that explains the permeation process for both organic and inorganic species in devices such as permeation tubes must consider such correlations.

The rate of permeation of the species is dependent not only on the gradient of its own chemical potential (that is, concentration) but, also upon the gradient of the chemical potential of other components, as well as other external cross-coupled factors (that is, stress, strain, electric field, polarization, charge, voltage, temperature, and entropy).

For a polymer-gas permeation device near equilibrium, the interdependences of the external factors just cited are assumed to be linear. These factors and their resultant cross-coupled interactions, which have been neglected or ignored in past studies of permeation through polymer substrates, are considered to be negligible for certain devices [13]. Depending upon the devices used, some of these interactions may need to be included to obtain a more complete understanding of their diffusion, permeation, and solubility phenomena. A comprehensive but fundamental approach to the general problem of permeation devices may be afforded by a more detailed treatment that explores the cross-coupled relationships between the permeator and the polymeric substrate, with specific attention given to the factors of stress, strain, electric field, polarization, charge, voltage temperature, heat capacity, and entropy. Such a treatment is beyond the scope of this present consideration. However, operating within the physical constraints of the permeation device, any of the previously mentioned calibration systems may be used with a high level of confidence.

COGAS System

To meet the ever increasing need for gaseous Standard Reference Materials (SRMs) the National Bureau of Standards has developed a Computer Operated Gas Analysis System

(COGAS) for calibration of both high pressure gas mixtures in cylinders and for sulfur dioxide (SO_2) permeation devices. For the calibration of SO_2 permeation devices, a considerable effort was expended to develop a system that reduced the amount of time and labor required for calibration while at the same time increasing the precision of measurement. The system operates automatically by sampling in sequence any number of permeation device chambers, up to 27. Each chamber is held in a temperature-controlled bath and contains a single SO_2 permeation device. The sample gas from the chamber is routed to an appropriate analytical instrument that is calibrated by permeation devices that have been calibrated gravimetrically. The data are collected, stored, and automatically processed to give a permeation rate for the SO_2 permeation devices measured.

Description of COGAS System

Basically, COGAS (Fig. 1) consists of 30 chambers immersed in a controlled temperature bath, each of which holds a single SO_2 permeation device. Dry air equilibrated to the temperature of the bath flows across each device at a rate of 100 mL/min. The air exiting the chambers is vented, except when a particular chamber is in use as a sample chamber. When a sample is measured, the air flow is routed through a flow controller, a selected manifold (Manifold 2 in Fig. 1), and a solenoid valve corresponding to the chamber of interest. The regulating valve is used to mix the SO_2-rich air leaving the chamber with fresh dry air, still maintaining the previously established flow rate of 100 mL/min. The sample gas is then routed to the mixing chamber before going to the analytical instrument. Between sample analysis and before the first sample is taken, the flow system is evacuated through the vacuum line as indicated in Fig. 1. The sequence of operations is controlled by a microcomputer that determines also the length of time of evacuation, sampling time, and collects, stores, and processes the data.

Specific COGAS Components

The entire system is constructed of commercially available components and materials. All of the components of the system that are in contact with the gas to be analyzed (that is,

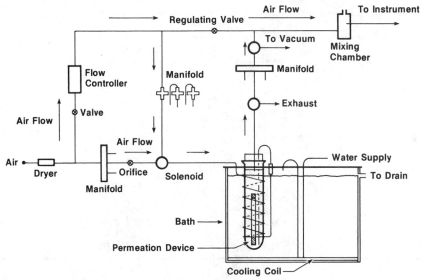

FIG. 1—*Schematic of COGAS system.*

SO_2) are made of stainless steel, with the exception of the solenoid valve seats and the glass chambers containing the devices. The solenoid valves are operated through solid-state relays (not shown), and each is equipped with a pilot light to indicate its energized condition.

The heart of the system is a commercially available microcomputer. A variety of such devices are available today, making the adaptation of a system such as COGAS to almost any sequential analysis scheme rather easy. The microcomputer should have the capability for operating a number of independent circuits, relays, solenoids, and for reading the output of various supporting instrumentation. The key control element of COGAS is the digital voltameter that serves as an analog-to-digital converter for the analytical instrument.

A variety of software support has been written to accommodate the various operating patterns and data handling. However, it should be stressed that COGAS is flexible enough that any combination of components (hardware, software, and system control) that satisfy the control and data processing requirements will be satisfactory, provided that the final results of the measurements are valid (see Table 2).

Sequence of Operation

Typically, 27 permeation devices and three standards are studied at any one time. The three standards used as references are gravimetrically calibrated and are not of the same manufactured batch. When the calibration is started, the entire system is evacuated. The solenoid valves are then closed and the analytical cycle begins. First, the system is evacuated after which the solenoid valve to the first reference chamber is opened and flow is established to the analytical instrument. Flow is continued through the instrument for a predetermined length of time. The time is determined by knowledge of the instrument response time and the approximate concentration being analyzed. The signal from the analyzer is not read or stored by the microcomputer until a sufficient length of time has elapsed for equilibrium to be established. The equilibration time is determined from an analog plot of the analyzer signal as a function of time. After this time period, the instrument signal is read and the average value is computed and stored. At this time, the solenoid valve is closed to the reference chamber. The system is then evacuated before the solenoid valve to the first sample chamber is opened. Each sample chamber is stepped through in turn, with a rede-

TABLE 2—*Comparison of gravimetric and COGAS permeation rates[a] for 5 and 10 cm SO_2 permeation tubes.*

Sample	Gravimetric Rate	COGAS Rate	Δ%
	5 CM		
A	1.33 ± 0.01	1.32 ± 0.01	−0.7
B	1.27 ± 0.01	1.26 ± 0.01	−0.8
C	1.35 ± 0.01	1.34 ± 0.01	−0.7
D	1.30 ± 0.01	1.29 ± 0.01	−0.8
E	1.25 ± 0.01	1.25 ± 0.01	0.0
F	1.28 ± 0.01	1.27 ± 0.01	−0.8
	10 CM		
G	2.51 ± 0.03	2.53 ± 0.03	0.8
H	2.64 ± 0.03	2.64 ± 0.03	0.0
I	2.59 ± 0.03	2.60 ± 0.03	0.4
J	2.54 ± 0.03	2.55 ± 0.03	0.4
K	2.53 ± 0.03	2.54 ± 0.03	0.8
L	2.61 ± 0.03	2.62 ± 0.03	0.7

[a] Rate = μg/min.

termination of the reference chamber after every five sample readings. Between each measurement, the system is evacuated for a predetermined length of time. After the 27 samples have been compared to Standard 1 (in Reference Chamber 1) in the manner described, the same process is repeated using Standards 2 and 3. The use of a complete set of data from any one device has served admirably as a permeation device calibration system. Table 2 shows the results of a comparison of the gravimetric rates and COGAS rates for 5 and 10 cm SO_2 permeation devices. With the exception of Samples E and H, all of the devices studied show a difference in rate of less than 1%.

Conclusions

For typical air pollutants such as NO_2 and SO_2, permeation device technology has developed to a state where reliable, stable devices are routinely fabricated. As new materials and applications are explored and developed, a more detailed and fundamental understanding of diffusion, solubility, and permeation is needed to explore the effect of nontraditional parameters.

The development of a calibration system, COGAS, allows for automatic analysis of a large number of permeation devices relative to known gravimetric standards. The entire operation consists of an intercomparison with three gravimetrically calibrated standards. For statistical reasons, the intercomparisons are performed seven times, producing 21 individual data points for each sample. Internal checks are made of each of the standards that are not being used as a reference, by measuring them as if they were a sample. This gives an immediate internal check as to how well the system is performing.

The COGAS system has been used successfully for the certification analysis of thousands of samples, including both permeation devices and compressed gas mixtures in cylinders. The system is controlled by a microcomputer, that is also used to process and store the data. The marked advantages of such a system are:

1. Improved precision of intercomparison.
2. Reduced manpower requirements.
3. Versatility in analytical sequences.
4. Adaptability to the analysis of many gaseous SRMs.

Because COGAS is essentially a dynamic dilution calibration system, the physical precautions outlined earlier also must be considered. The manner in which COGAS is operated clearly tests and monitors the system's adherence to sound functioning parameters. This is demonstrated by evaluating data such as those given in Table 2. Specifically, when used as a calibration system for SO_2 permeation devices, COGAS is constantly monitored through the results obtained when analyzing a gravimetric standard as if it were a test sample. The results illustrated in Table 2 provide a strong indication of how well COGAS behaves as a trace calibration system for permeation devices, and provides a confirmation of how well the systems parameters (physical precautions) are under control.

References

[1] Hersch, P. A., *Analytical Chemistry*, Vol. 32, 1960, p. 1030.
[2] O'Keeffe, A. E. and Ottman, G. C., *Analytical Chemistry*, Vol. 38, 1960, p. 760.
[3] Hughes, E. E., Rook, H. L., and Deardoff, E. R., *Analytical Chemistry*, Vol. 49, 1977, p. 1823.
[4] Lodge, J. P., "Production of Controlled Test Atmospheres," *Air Pollution*, A. C. Stern, Ed., Academic Press, New York, 1962.

[5] Hodgeson, J. A., Baumgardner, R. E., Martin, B. E., and Rehme, K. A., *Analytical Chemistry*, Vol. 43, 1971, p. 1123.
[6] Fried, A. and Hodgeson, J., *Analytical Chemistry*, Vol. 54, 1982, p. 278.
[7] Stern, A. C., *Air Pollution-Analysis, Monitoring and Surveying*, Vol. II, Parts I and VI, 2nd ed., Academic Press, New York, 1968.
[8] Bamesberger, W. L. and Adams, D. E., *Environmental Science and Technology*, Vol. 3, 1969, p. 258.
[9] Michaels, A. S. and Bixler, H. J., *Journal of Polymer Science*, Vol. 50, 1961, p. 413.
[10] Van Amerenger, G. J., *Journal of Applied Physics*, Vol. 17, 1946, p. 972.
[11] Michaels, A. S. and Parker, R. B., *Journal of Polymer Science*, Vol. 41, 1959, p. 53.
[12] Michaels, A. S. and Bixler, H. J., *Journal of Polymer Science*, Vol. 50, 1961, p. 391.
[13] *Physics and Chemistry of Organic Solid State*, Vol. II, D. Fox, M. M. Lebes, and A. Weissberger, Eds., Interscience, New York, 1965, Chapter 6.

Alan Fried[1] and Robert Sams[1]

Tunable Diode Laser Absorption Spectrometry for Ultra-Trace Measurement and Calibration of Atmospheric Constituents

REFERENCE: Fried, A. and Sams, R., **"Tunable Diode Laser Absorption Spectrometry for Ultra-Trace Measurement and Calibration of Atmospheric Constituents,"** *Sampling and Calibration for Atmospheric Measurements, ASTM STP 957,* J. K. Taylor, Ed., American Society for Testing and Materials, Philadelphia, 1987, pp. 121–131.

ABSTRACT: There has been an ongoing quest for development of ever more sensitive and selective detection methods for studying various gas molecules of atmospheric importance. Both laboratory and ambient studies often require instrumentation capable of measuring ultra-trace concentrations of such gases at, and below, the parts-per-billion range. Concurrently, accurate calibration standards, particularly those verified by independent techniques, are also required. The sensitive, selective, and versatile technique of infrared tunable diode laser absorption spectrometry is especially well suited for ultra-trace gas measurements and calibration standards verification. This combination, which is not shared by many other measurement techniques, results from the fact that diode laser spectrometers can be operated in an absolute mode as well as a more sensitive relative mode. In this paper, we discuss these capabilities and present specific examples for the measurement and calibration of ultra-trace levels of the important atmospheric gases nitrogen dioxide (NO_2) and hydrochloric acid (HCl). In addition, we further discuss the application of tunable diode laser absorption spectrometry in the verification of NO_2 permeation standards using the gas-phase titration reaction between nitric oxide (NO) and ozone (O_3).

KEY WORDS: air quality, calibration, sampling, atmospheric measurements, laser absorption spectrometry, ultra-trace measurements

It is well known that trace atmospheric gases, resulting directly or indirectly from man's activities, often dictate and control the chemistry of our atmosphere. Several well-publicized environmental consequences of man's impact include the "greenhouse effect," stratospheric depletion of ozone, and the occurrence of "acid rain." A comprehensive understanding of these as well as other perturbations to the atmosphere requires both laboratory and ambient studies of a large number of atmospheric gases. Such studies in turn, rely on instrumentation capable of selectively measuring ultra-trace concentrations of these atmospheric constituents at, and below, the parts-per-billion (ppb) range.

In addition to highly sensitive measurement techniques, atmospheric-related studies also require accurate calibration procedures and standards in the ppb concentration range. For the most reliable results, it is essential that the accuracy of the calibration procedures and standards be verified by independent techniques based upon different measurement principles. Many sensitive measurement techniques, however, exhibit a relative response, and,

[1] Group leader and research scientist, respectively, Analytical Laser Spectrometry Group, Center for Analytical Chemistry, National Bureau of Standards, Gaithersburg, MD 20899. Dr. Fried is presently with the National Center for Atmospheric Research, P.O. 3000, Boulder, CO 80307.

thus under most circumstances, the concentration for the particular gas standard of interest can only be verified in the ppb range by calibrating against accurately prepared gas standards at higher concentrations. The ultimate accuracy in the ppb range will obviously depend upon the accuracy of the higher concentration gas standard as well as the dilution or extrapolation procedure used in the intercomparison. While this can in principle yield accurate calibration standards in the ppb range, it is not based upon a totally independent verification procedure. Absorption spectrometry, on the other hand, can provide an absolute and independent concentration determination for most gas standards of interest. In this paper, we discuss the application of absorption spectrometry, more specifically infrared tunable diode laser absorption spectrometry, in the measurement and verification of ultra-trace gas standards for atmospheric constituents.

Infrared Tunable Diode Laser Absorption Spectrometry

Measurement techniques based upon absorption of infrared radiation (3 to 30 μm spectral region) have proven to be extremely attractive in atmospheric studies. Most trace gases of atmospheric importance exhibit moderately strong absorption in this spectral region while the major constituents, oxygen (O_2) and nitrogen (N_2) do not. Furthermore, absorption lines in this spectral region, which result from vibrational-rotational transitions, predominantly appear as sharp discrete features for small molecules generally containing less than five atoms. This is frequently the case when the sampling pressure is in the 0.13 to 6.7 kPa range (1 to 50 torr) and the infrared radiation source is sufficiently narrow in linewidth. Under such conditions, the spacing between individual absorption features generally exceeds typical absorption linewidths, which for most small atmospheric molecules is in the 0.001 cm^{-1} range. In some cases, molecules containing more than five atoms may also display spectral regions where the rotational features are well resolved.

Tunable diode lasers (TDLs) are extremely attractive sources of infrared radiation. These lasers emit infrared radiation in very narrow linewidths, oftentimes much less than 10% of the linewidth of typical absorption features. This is essential in achieving the high resolution inherent to low pressure infrared measurements just discussed. The resulting high resolution spectral features readily allow line-fitting analysis to be applied. As will be discussed shortly, rare overlapping spectral features will generally stand out employing such analysis, thus ensuring the utmost in specificity. Tunable diode lasers, in addition, exhibit a rather broad frequency coverage spanning the entire near infrared region, and can be continuously tuned over small frequency intervals. All of these attributes are directly responsible for the high sensitivity, selectivity, and versatility demonstrated in many TDL studies of atmospheric molecules [1–4].

In tunable diode laser absorption spectrometry (TDLAS), there are two distinctly different modes in which to acquire data: direct absorption and second harmonic detection. In direct absorption, the laser is scanned through spectral absorption features associated with the particular species under study and the incident (I_0) and transmitted (I) intensities are recorded. As shown in Fig. 1, these data, together with measurements of the sample pressure (P), pathlength (L), and line center absorption coefficient ($\alpha(v_0)$) are used in the well-established Beers-Lambert absorption expression to determine the concentration (C) of the absorber species. This data acquisition mode yields absolute concentration determinations solely based upon photometric and molecular parameter measurements. If care is exercised to eliminate various potential sources of systematic errors in the photometric measurements [5], TDL direct absorption can yield concentration determinations for various gas standards with absolute accuracies in the 1 to 10% range. Minimum detectable absorbances, ln (I_0/I), around 10^{-4} can be achieved using this data acquisition mode. Employing an absorption

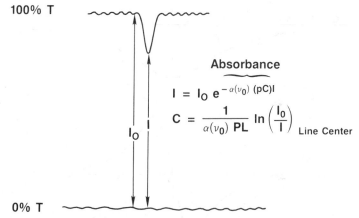

100% T

0% T

$$I = I_0\, e^{-\alpha(\nu_0)\,(pC)l}$$

Absorbance

$$C = \frac{1}{\alpha(\nu_0)\,PL}\, \ln\!\left(\frac{I_0}{I}\right)$$

Line Center

I_0

I

FIG. 1—*Illustration of the direct absorption technique in the absolute determination of gas concentrations* (C) *employing Beers-Lambert law. The incident* (I_0) *and transmitted* (I) *intensities are recorded as the laser wavelength is scanned through an absorption feature for the molecule of interest. The line center absorption coefficient* ($\alpha(\nu_0)$), *pressure* (P), *and pathlength* (L) *are also required in this determination.*

pathlength of 40 m, this corresponds to a minimum detectable concentration in the ppb range for most moderately strong infrared absorbing gases. Important atmospheric molecules such as nitric oxide (NO), nitrogen dioxide (NO_2), and nitric acid (HNO_3), for example, can be quantitatively measured at concentrations around 30 ppb with a precision of 10%. Even lower concentrations or higher precisions or both can be achieved using longer optical pathlengths.

In addition to direct absorption, TDL measurements can also be carried out using the technique of second harmonic (second derivative) detection. In this mode, measurements are carried out by simultaneously superimposing an external modulation waveform (in the kilohertz (kHz) frequency domain) on the diode laser tuning current. A lock-in amplifier referenced to this frequency is used to synchronously detect the second harmonic signal of the frequency-modulated absorption that occurs at twice the modulation frequency. The direct, first, and second harmonic absorption lineshapes are all shown schematically in Fig. 2. The harmonic lineshapes are similar in appearance to the profiles obtained by taking appropriate derivatives of the direct absorption. For sufficiently small modulation amplitudes, true derivative profiles are obtained.

Second harmonic detection eliminates the necessity of measuring small differences between two large intensities, I and I_0, as is the case for direct absorption of trace atmospheric constituents. Further advantages of second harmonic detection over direct absorption are also realized: signal processing at twice the modulation frequency (1) eliminates a strongly sloping background often present in direct absorption, (2) produces a zero baseline (as shown in Fig. 2), (3) minimizes the vulnerability to small drifts in the laser output power, (4) reduces low frequency noise due to the kHz detection regime, and (5) displays a large discrimination against signals that do not have a strong wavelength dependence. All of these factors result in an increased detection sensitivity over direct absorption. Minimum detectable absorbances achieved in various TDL studies employing second harmonic detection have consistently been in the 10^{-5} range. This factor of ten improvement enables most atmospheric constituents of interest to be detected at sub-ppb concentration levels using moderate absorption pathlengths around 40 m. At the National Bureau of Standards (NBS), recent experimental results suggest that this detection limit can be further improved by an

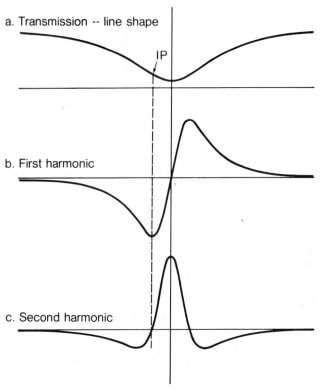

FIG. 2—*Direct absorption (transmission), first and second harmonic lineshapes. The point marked as IP is the inflection point indicating the maximum negative slope in the direct absorption profile.*

additional factor of ten utilizing the technique of sweep integration [6] coupled with second harmonic detection.

The increased sensitivity of second harmonic detection, however, is gained at the expense of a relative response. Unlike direct absorption, many instrument and experiment-dependent factors must be taken into account when attempting to deduce absolute concentrations from the measured second harmonic response. As a result, accurate quantitative analysis by second harmonic detection requires, for all practical purposes, calibration standards. Thus, second harmonic detection cannot by itself be used as an independent means of verifying trace gas standards. However, as will be discussed shortly, this method of detection can be combined with direct absorption to achieve the advantages inherent in both techniques: the high sensitivity of second harmonic detection coupled with the absolute response of direct absorption.

Tunable Diode Laser Apparatus

A typical TDL optical set-up for detection of trace atmospheric gases is schematically shown in Fig. 3. The output of a TDL, mounted in a cryogenic cold head assembly, is collimated and subsequently split into two paths by a beam splitter, as shown. Approximately 10% of the beam is reflected and is either directed through a wavelength reference cell, containing a high concentration of the sample gas, or an interferometer. The former is used as a convenient means of tuning the laser to transitions associated with the sample gas, while the latter generates high precision and accurate relative frequency markers necessary for accurate absolute direct absorption measurements. In either case, the beam is ultimately focused onto a mercury cadmium telluride infrared detector, Detector (B), shown in Fig. 3.

The main beam, transmitted by the beam splitter, is directed through a concentration reference cell and focused into a multi-pass White cell by a series of flat and off-axis parabolic mirrors. As will be discussed, this concentration reference cell, which is also filled with a high concentration of the sample gas, is used for calibration. The beam emerging from the White cell (the sample cell) is then focused onto infrared Detector (A). In our laboratory, we typically adjust the White cell optics to achieve 20 traversals of the diode beam in a 2-m base path cell to effect a total optical pathlength of 40 m. With proper optical design, pathlengths of several hundred meters can be achieved.

Numerous modifications of the setup shown in Fig. 3 have been employed in various TDL studies. Simultaneous lasing in many different diode modes, for example, often requires a monochromator to be added to the optical layout for mode selection. As discussed in Refs 5 and 6, additional modifications are often necessary to minimize systematic errors in the direct absorption data acquisition mode.

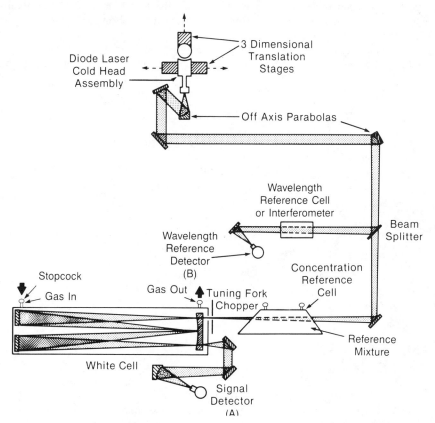

FIG. 3—*Schematic of typical tunable diode laser experimental setup.*

Measurement Selectivity

Because of the high resolution inherent in TDLAS, it is rare that one encounters spectral interferences. However, even when they do arise the high resolution enables line-fitting analysis to be applied as an additional tool to further ensure high selectivity. The power of this procedure is demonstrated in Fig. 4 where two different absorption lines of NO_2 in the 1605 cm^{-1} region have been recorded at very low pressures around 30 μm. Under these conditions, both line profiles should be accurately described by Gaussian functions. The halfwidths at halfheight of the corresponding peak absorbances can be accurately calculated employing the well-known Doppler relationship [7]. Over small frequency intervals, as is certainly the case here (the line separation is 0.022 cm^{-1}), these halfwidths are, for all practical purposes, equivalent. Thus, identical Gaussian functions, except for differences in the integrated strengths, were employed in fitting both lines.

In the upper portion of Fig. 4, we display both observed and calculated (fit) transmission line profiles for the two lines of NO_2. The bottom profile displays the observed minus calculated residuals. As can be seen, the fit for the second line (designated $9_{4,6+}$) is excellent while the fit for the first line (designated $9_{4,6-}$) is very poor. Both the transmission and residual profiles clearly indicate an unresolved weaker third line on the right shoulder of the first line. The observed transmission in this shoulder region is much less than that calculated for a single Gaussian line profile. This gives rise to the very large negative residual displayed in the lower trace of Fig. 4. At line center, the reverse is true. This is because the halfwidth is fixed and thus the extra absorption due to the unresolved third line becomes prevalent at line center.

In this particular case, the unresolved extra line also happens to be due to NO_2, and we knew of its presence beforehand from line assignments. This spectral region, in fact, was

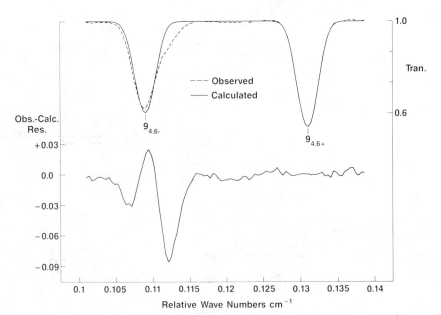

FIG. 4—(Upper Trace) *Observed and calculated (fit) transmission spectrum assuming two Doppler-limited NO_2 lines at 1605.5 cm^{-1}. (Lower Trace) Observed minus calculated residuals for the above spectrum. The poor fit of the first line indicates an unresolved additional line, as discussed in the text.*

chosen because it illustrates quite graphically the power of line-fitting analysis in distinguishing between lines unperturbed and perturbed by spectral interferences. Unresolved and unsuspecting interferences associated with gases other than the analyte being studied would display behavior similar to that shown in Fig. 4. In the most severe case, however, the interferant and analyte lines could exactly coincide. Under such circumstances, which are indeed very rare, analysis of several lines would be required to reveal the presence of the interferant. In addition to the low pressure direct absorption scans of Fig. 4, equivalent line-fitting analysis could also be applied to studies carried out at higher pressures as well as second harmonic measurements. The latter, however, requires modeling the second harmonic response function [1,8] or, as in the present case, two or more similar lines for comparison purposes.

The very high inherent spectral resolution coupled with the discriminatory nature of line-fitting analysis are significant attributes of TDLAS not common to many other techniques. As a result, one is not required to a priori test for all possible known interferences before carrying out trace measurements. Even when all suspected interferences in other detection techniques are tested, there is still the possibility of unknown interferences.

Application of TDLAS for Measurement and Verification of Calibration Standards for Atmospheric Constituents

Absolute Concentration Determination Methods

As we have discussed previously, direct absorption can be utilized to quantitatively measure, with a precision around 10%, moderately strong infrared absorbers at concentrations around 30 ppb in pathlengths of 40 m. A factor of five increase in the optical pathlength would allow a factor of five lower sample concentration to be detected with the same precision. This of course is predicated on the fact that the noise amplitude remains constant as the pathlength is increased, a situation not always easily achievable.

However, by coupling second harmonic detection with direct absorption, absolute concentration determinations can be carried out in the ppb and sub-ppb concentration range of interest for standards verification. This is achieved by ratioing the second harmonic response measured for the sample gas flowing through the White cell to that of a reference gas, the concentration of which is determined by direct absorption. In this approach, three different schemes can be utilized. Each scheme differs in the placement, composition, and concentration of the reference gas. In all three, however, the concentration of the sample gas (C_s) flowing through the White cell is determined using the following basic relationship

$$C_s = C_r \frac{S_s (I_0)_r L_r (RF)_r}{S_r (I_0)_s L_s (RF)_s} \qquad (1)$$

In this expression, C, S, I_0, L, and RF represent the concentrations, second harmonic signals, incident intensities, pathlengths, and response factors, respectively, for the sample gas, s, and the reference gas, r.

In the first calibration scheme, the reference and sample gases are identical, as are the temperature and total pressure of both gases. The same is also true in the second calibration scheme. However, in the first approach, the reference gas is contained or continuously passed through a different (short pathlength) cell than the sample gas flowing through the White cell. This concentration reference cell, as shown in Fig. 3, contains a reference mixture many times higher in concentration than the sample gas (typically 1000 times) so as to achieve an optical absorbance in the 0.01 to 0.25 range. Smaller absorbances will generally degrade the precision with which the reference gas can be measured by direct absorption.

Much larger absorbances, on the other hand, may cause a nonlinear second harmonic response in the reference mixture.

For sample measurements, the high concentration reference cell must either be removed or evacuated. The former procedure, however, requires a measurement of the incident intensity both with and without this reference cell in place. Because of the attendant uncertainty this introduces, evacuation of the reference cell in place is preferred. The intensity will thus remain constant during sampling and calibration phases and hence cancels out of Eq. 1. Since the temperature, pressure, and composition of the sample and reference gases are also equivalent, the same is true of the response factors in Eq. 1. This is also the case for the second calibration scheme, as will now be discussed.

The second calibration approach is very similar to that of the first. The only difference is that the reference mixture is passed through the White cell and not a separate smaller pathlength cell. As a result, the pathlength ratio, L_r/L_s, in Eq. 1 is constant and also cancels out. In this procedure, only marginally higher reference gas concentrations than the sample gas are required. In measuring sample gas concentrations of NO_2 in the 1 ppb range in a 40-m White cell, for example, reference mixture concentrations in the 30 to 100 ppb range are required for quantitation by direct absorption, as discussed previously. Much higher concentrations, particularly for NO_2 as well as other similar highly polar reactive gases, may contaminate the sampling cell. As we often experience, once the sample cell is exposed it becomes very difficult to remeasure ppb concentrations of the sample gas without experiencing undue problems caused by outgassing of the reference mixture from the cell walls.

In the third calibration scheme, an approach developed at NBS [1] and presently being employed in field studies [9], the reference mixture is no longer identical to the sample gas. Instead, a nonreactive stable gas, possessing absorption features in the same spectral region as the sample gas, is passed through the sample cell. In our studies of the reactive gas, hydrochloric acid (HCl) [1], a stable mixture of CH_4 was employed as the reference gas. As in the second approach, no additional cell is required for calibration and thus the intensity and pathlength ratios cancel out of Eq. 1. This technique is beneficial because the concentration of the reference mixture can be adjusted to closely match second harmonic signals for the sample and reference gases. Thus, large signal extrapolations are eliminated along with any concerns of sample cell contamination.

This third calibration approach is also advantageous for atmospheric TDL measurements where reactive gas calibration systems may be difficult or next to impossible to deploy. Furthermore, by measuring ambient concentrations of the reference mixture, by gas chromatography, for example, open frame White cell systems can also be calibrated using this technique.

However, because of the different composition of the sample and reference gases, the third calibration approach does require an additional consideration. In contrast to the first two procedures, the lineshapes for the sample and reference gases are rarely the same. As a result, the response factors for the two gases can be dramatically different. As shown in Eq. 1, the response factor ratio becomes an integral part in determining the sample gas concentration. This ratio, as shown in our HCl studies [1], can be accurately determined by modeling the second harmonic response or by carrying out empirical measurements or both on known concentrations of both gases.

Standards Verification by TDLAS

The redundancy in calibration afforded by TDLAS as just discussed is a significant advantage of the technique for standards verification of reactive gases. This redundancy has in fact been exploited in measuring sample gas concentrations in the 1 ppb range for the important atmospheric molecule HC1. A 40-m White cell was also employed as the sample cell throughout these measurements. A summary of the results of this study are shown in Fig. 5. For a thorough description of the experimental details and the results, the reader is referred to Ref *1*. In Fig. 5, we display the second harmonic signal resulting from various trace HCl concentrations, generated from a commercial permeation calibration system, as a function of concentration based upon four independent calibration techniques. The response curve indicated by [HCl]^{permeation} is based on the calculated permeation concentration, the same approach that normally would be taken with other trace gas standards. The response curve and the single point determinations designated, [HCl]^{IR} and [HCl]^{CH4}, are calibrations based upon the first and third calibration approaches, respectively, which have

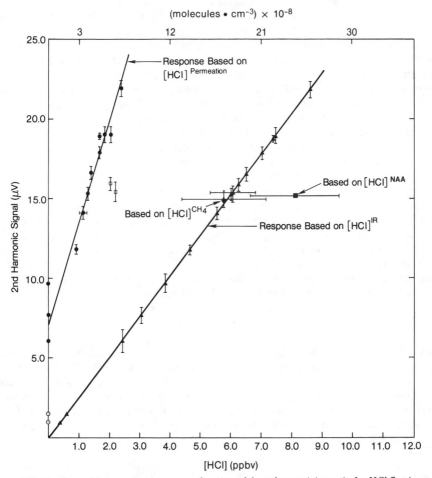

FIG. 5—*Second harmonic response as a function of the volume mixing ratio for HCl flowing through a White cell at 8.9 torr based upon four independent determinations. The response curve and single point determinations, [HCl]^{IR} and [HCl]^{CH4}, are based on IR ratioing techniques using HCl and CH_4, respectively, as the reference gas. The point [HCl]^{NAA} is based on the quantitative trapping of HCl, at the White cell inlet, on nylon filters followed by neutron activation analysis for Cl. The [HCl]^{permeation} response curve is based on the calibrated permeation rate of a commercial HCl permeation tube. Refer to the text and Ref 1 for more thorough details.*

been discussed previously. The additional calibration point, designated $[HCl]^{NAA}$, was obtained by quantitatively absorbing HCl at the White cell inlet on ultrapure nylon filters followed by subsequent neutron activation analysis of the trapped acid-chloride. The reader is referred to Ref *1* for details on how this additional calibration point was incorporated in our intercomparison.

As can be seen, the $[HCl]^{IR}$, $[HCl]^{CH_4}$, and $[HCl]^{NAA}$ calibrations are all within the mutual uncertainties (one sigma) estimated for each technique. This firmly establishes the concentration of the permeant sample gas flowing through the White cell. The fourth calibration based upon the known permeation emission rate, however, is clearly outside the range of statistical agreement. This indicates a fundamental problem with the permeation source. If trusted without further verification, one would erroneously obtain a factor of 2.5 lower HCl concentration than actually present. Additional data similar to Fig. 5 and subsequent recalibration of the permeation device did, in fact, reveal the source of the problem; the seal on the Teflon permeation membrane was no longer intact, and thus the emission rate unknowingly increased during these intercomparisons. Similar TDL measurement redundancies can readily be applied in the verification of many other permeation as well as static trace gas standards. While permeation sources are inherently accurate devices, many potential systematic errors can still arise, as just shown. Since permeation sources are often the only available means of generating trace concentrations of various atmospheric constituents, such redundancy becomes particularly important in atmospheric studies.

At NBS, we have expanded our TDL research efforts to provide unequivocal accuracy in NO_2 gas standards, both in compressed gas as well as in trace permeation gas standards. As an important part of this program, we are presently involved in an effort to improve NO_2 absorption and pressure-broadening coefficients upon which the accuracy of direct absorption TDL measurements ultimately depend. To accomplish this, direct absorption measurements will be carried out on accurately known concentrations of NO_2 in the ppm range generated by permeation sources. Since the goal of this work is to improve the uncertainty of the molecular parameters to better than the present 10% range, the TDL procedures described previously obviously cannot be employed as an independent means of verifying the permeation rates.

Conveniently, the NO_2 permeation standards can be independently verified employing gas-phase-titration (GPT) involving the rapid bimolecular reaction

$$NO + O_3 \rightarrow NO_2 + O_2$$

In an extensive study of this reaction, the NO and ozone (O_3) concentrations consumed were found by Fried and Hodgeson [10] to be in agreement to within 4%. Agreement to better than 1% was obtained when NO concentrations were intercompared with NO_2 concentrations produced from this reaction. Further studies of this reaction are planned to reduce or explain, or both, the small apparent O_3 bias. However, even at this current stage, GPT in conjunction with permeation measurements will enable us to accurately carry out NO_2 absorption coefficient measurements.

Summary

Tunable diode laser absorption spectrometry has been shown to be an extremely versatile, sensitive, and selective technique for ultra-trace measurements of numerous atmospheric constituents in both laboratory and ambient studies. Furthermore, by coupling the absolute response of direct absorption with the more sensitive second harmonic method of detection, we have shown how tunable diode laser absorption spectrometry is especially well suited for standards verification in the ppb concentration range. Such standards verification, more-over, can be carried out employing multiple calibration approaches. As we have observed in previous studies of trace reactive gases, this redundancy is of utmost importance in verifying the accuracy of ultra-trace gas standards for atmospheric constituents. Because of these capabilities, the role of tunable diode laser absorption spectrometry will continue to expand in all aspects of atmospheric measurements.

References

[1] Fried, A., Sams, R., and Berg, W. W., *Applied Optics*, Vol. 23, 1984, p. 1867.
[2] Hastie, D. R., Mackay, G. I., Iguchi, T., Ridley, B. A., and Schiff, H. I. *Environmental Science and Technology*, Vol. 17, 1983, p. 352A.
[3] Reid, J., Garside, B. K., Shewchun, J., El-Sherbiny, M., and Ballik, E. A. *Applied Optics*, Vol. 17, 1978, p. 1806.
[4] Pokrowsky, P. and Herrmann, W., *Proceedings*, Society of Photo-Optical Instrumentation Engineers, Vol. 286, 1981, p. 33.
[5] Sams, R. and Fried, A., *Journal of Applied Spectroscopy*, Vol. 40, 1986, p. 24.
[6] Jennings, D. E., *Applied Optics*, Vol. 19, 1980, p. 2695.
[7] Mavrodineanu, R. and Boiteux, H. in *Flame Spectroscopy*, Wiley, New York, 1965, p. 520.
[8] Reid, J. and Labrie, D., *Applied Physics B*, Vol. 26, 1981, p. 203.
[9] Restelli, G. and Cappellani, F., *Applied Optics*, Vol. 24, 1985, p. 2480.
[10] Fried, A. and Hodgeson, J., *Analytical Chemistry*, Vol. 54, 1982, p. 278.

W. D. Dorko[1] and E. E. Hughes[1]

Special Calibration Systems for Reactive Gases and Other Difficult Measurements

REFERENCE: Dorko, W. D. and Hughes, E. E., **"Special Calibration Systems for Reactive Gases and Other Difficult Measurements,"** *Sampling and Calibration for Atmospheric Measurements, ASTM STP 957,* J. K. Taylor, Ed., American Society for Testing and Materials, Philadelphia, 1987, pp. 132–137.

ABSTRACT: The most popular method for determining a component of interest in a gas mixture is to use a detector that responds to the analyte and is calibrated directly with a cylinder gas mixture containing the analyte at the approximate concentration of interest. In many instances, however, this cannot be achieved; cylinder calibration mixtures may be unstable or the direct response detector may not be sensitive enough. The unstable cylinder mixture can be replaced by a dynamic dilution gas calibration system where the analyte can be introduced by any one of several means including permeation devices, or a higher concentration gas mixture that can be more easily stabilized. If the detector for the analyte is insensitive or difficult to calibrate, then the analyte can be converted to another compound for which there is a detector that is sufficiently sensitive and easy to calibrate.

In the case of analytical systems, such as gas chromatography, where separation occurs, a mixture of a stable gas can be used to calibrate for a reactive gas if the relative detector responses are known. There is no one particular way to either provide calibration mixtures for the analysis of all reactive gases or to provide a means for the determination of difficult compounds. Any system that will produce a mixture of the required concentration, of acceptable uncertainty, and of stability sufficient for the analysis at hand can be a special calibration system.

KEY WORDS: calibration systems, gas mixtures, reactive gases, calibration standards

The most popular method for providing calibration gas standards is to use a cylinder mixture of gases. The mixture required comes directly from the cylinder. However, for reactive gases at low concentrations, different approaches may be required. Two useful methods for generating calibration mixtures of reactive gases are by means of a permeation tube system or a dynamic dilution system. There can be a variety of inputs to a dynamic dilution system including a cylinder mixture of a high concentration reactive gas that is then diluted to the appropriate level. In most situations, either one or a combination of both, a permeation system and dilution system will yield a solution to a problem. However, as a preliminary to discussing a few specific cases, the term "difficult" gas should be addressed.

It is probably valid to assume that any gas mixture that has not been studied in detail will be "difficult." There could be wall adsorption or desorption, reaction with other species, or some other occurrence to change analyte concentration. The simple gases, carbon monoxide (CO), carbon dioxide (CO_2), propane (C_3H_8), methane (CH_4), hydrogen (H_2), and oxygen (O_2) have all been evaluated and their behavior in calibration mixtures, sampling systems and in analytical instrumentation is well understood. How then, would one approach the problem of preparing a calibration mixture of a substance whose behavior is unknown?

---

[1] National Bureau of Standards, Center for Analytical Chemistry, Gaithersburg, MD 20899.

A logical approach is to define the needs or requirements for the mixture. Can a mixture be prepared in a cylinder with adequate reliability for the eventual application? What problems will arise from physical and chemical reactions of the substance in the cylinder in addition to those involved in measurements of the quantity of the substance.

There is no one unique way to solve all difficult calibration problems. Instead of giving a bibliography of problem solutions, we will describe some tests that can be performed to recognize problems and give some examples of different means to provide characterized calibration mixtures or to perform difficult measurements.

Problem Recognition Tests

There are several tests that can be applied during the actual blending of a gas mixture that can give information concerning the behavior of the mixture and may allow one to decide whether to continue the particular process or to discard it and look for other means of preparing the blend.

Assume that a relative method of analysis exists for the species in question. (Obviously, if an absolute method exists, the problem is nonexistent.) One prepares a series of mixtures by one of the conventional blending techniques (which in the particular laboratory is known to be reliable) such as by gravimetry, pressure, or volumetric methods. The next step is to generate a signal with each blend in the series using an appropriate instrument or procedure, and to compare the signals. The results may be one of the three situations shown graphically in Fig. 1.

The results shown in Fig. 1A indicate that some adverse action, such as adsorption, may be occurring somewhere in the blending process, Fig. 1B may be due to the material of interest being present in the dilution gas, and Fig. 1C may be due to a faulty assay of the pure material of interest.

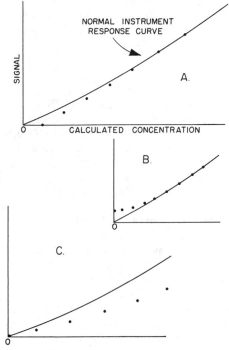

FIG. 1—*Analyte concentration of a gas cylinder, observed versus calculated.*

It should be understood that interpretation of this type of data requires understanding of the characteristic of any of the measuring devices used, as well as a familiarity with the behavior of gases in general. The results of an experiment such as this are not definitive but rather suggest a direction to follow.

If the results conform to Fig. 1A and it is suspected that adsorption is the cause of the difficulty, it can be confirmed by a simple experiment consisting of transferring some of the mixture in the suspect Cylinder 1 to another similar Cylinder 2. Cylinder 2 is originally an empty, clean cylinder treated as closely as possible as was Cylinder 1 that contains one of the mixtures in the series that showed a deviation from expected behavior.

If the concentration of the substance in Cylinder 1 (C_1) is measured just before transfer to Cylinder 2 and the concentration in Cylinder 2 (C_2) is measured after transfer, the two concentrations should be the same if no adsorption occurs. If C_2 is significantly less than C_1, then adsorption has occurred and it will be impossible to obtain a reliable concentration based on the mixing parameter alone.

A less direct confirmation that adsorption has occurred within a cylinder can be obtained by slowly reducing the pressure in the cylinder while monitoring the concentration. A significant rise in concentration indicates significant adsorption during the initial filling of the cylinder. However, a constant concentration during a slow reduction of cylinder pressure should not be taken to mean that no adsorption has occurred. Other factors such as the dryness of the sample and container or reducing the pressure in the cylinder too fast to achieve equilibrium may affect the behavior of the system. Keep in mind also, that a decision as to whether or not a particular gas mixture is stable should not be based on single results. No two gas cylinders are exactly alike, especially in regard to the condition of dryness of the interior, and any conclusion should be based on a composite picture generated by observations of many cylinders.

If the calibration curve generated from a set of mixtures indicates that there was an incorrect assay of the addition gas or contamination of the diluent gas, then the only recourse is to reanalyze. The intercept predicted from the curve is an indication of the magnitude of this problem and can be used as a guide to help you determine what to look for.

Dilution Systems

If now we have decided that all the evidence indicates we cannot prepare a predictable mixture in a cylinder, we look to the alternatives. The first choice is generally a dilution technique, such as low pressure static dilution, where a measured volume of an intermediate mixture is diluted in a container of known volume, or dynamic dilution, where a measured flowing stream of the intermediate mixture is mixed with a measured flowing stream of a diluent. The concentration of the intermediate mixture is at a level where the mixture is known to be stable or is measured just before use. These higher concentration intermediates are frequently much easier both to stabilize and to analyze by wet chemical techniques. For example, a dynamic dilution system was designed to generate low levels of ammonia (NH_3). The system consisted of dilution devices coupled with a wet chemical analytical unit to determine the high concentrations of NH_3. A high concentration mixture of ammonia was used as dilution input and this was analyzed by means of the analytical half of the system each day it was used. A similar system was designed for chlorine [1]. If the substance is so reactive that no stable intermediate level mixture can be prepared, then it is not a good candidate for dilution.

Dilution mixtures, either static or dynamic, are intended for immediate use and are not intended to be stored for long periods of time. Static dilution mixtures are generally less desirable than dynamic dilution mixtures because, while simple to prepare, they may degrade too rapidly to be useful.

Over the years, a number of devices other than the aforementioned intermediate mixtures have been developed for the introduction, at a controlled rate, of gas or vapor into a stream of diluent gas. These have depended on mechanical feeding, diffusion through capillary tubes, evaporation at known temperature (saturation), and permeation or effusive devices [2]. Each has had a particular application but, with the exception of permeation tubes, they have not been studied in sufficient detail or with a large enough variety of materials to allow any statement concerning their general applicability.

Permeation tubes are lengths of inert plastic tubing (usually Teflon™) sealed at both ends and containing the analyte of interest that is either a liquid or liquefiable gas. The analyte escapes by permeating through the tube wall at a constant, reproducible, temperature-dependent rate. Permeation tubes are held at a constant temperature in a carrier-gas stream of dry air or nitrogen to produce a gas mixture with the analyte concentration dependent on the permeation rate and the flow of the carrier gas as in ASTM Recommended Practice for Calibration Techniques Using Permeation Tubes (D 3609-79). Permeation tubes are available for a large number of substances including reactive and corrosive gases, such as sulfur dioxide (SO_2), nitrogen dioxide (NO_2), and hydrogen sulfide (H_2S).

Dynamic dilution systems may be evaluated in much the same way as a series of cylinder mixtures, that is, by observing the signal produced at a particular dilution and comparing it to the predicted value. The calibration of the dilution system can be accomplished quite easily using standard, stable gas mixtures monitored with a precise analytical method.

The predicted calibration curve may be slightly different from the observed curve, Fig. 2. This can occur because of the accumulation of uncertainties in the calibrations of the individual flow measuring devices; or the deviation might arise from interactions between the mixed flows that affect slightly the flow through the individual flow measuring devices [3]. Each flow device is calibrated separately keeping a known, constant downstream pressure. When the outputs of two or more flow devices are connected and the flows varied to produce a sequence of distinct concentrations, the downstream pressure may increase and might affect the calibrations, thus affecting the output concentration. Once the dilution system is calibrated, the unstable gas is substituted for the stable gas and the response at various dilutions is noted. Again, it is necessary to understand the analytical instrument used in order to be able to draw significant conclusions from the results. Some instruments have a linear response, signal versus concentration; while other instruments are slightly nonlinear; and still others have an exponential response. If a dilution system is used to generate standards for an instrument calibration and a nonlinear curve is observed, then a

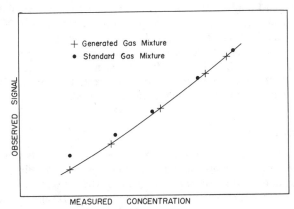

FIG. 2—*Predicted versus observed analyte concentration output of a dynamic dilution system.*

problem with the dilution system would be suspected if one did not know that the particular instrument response was indeed nonlinear.

Internal Standards

The time and effort necessary to evaluate a gas system for stability and reliability are enormous and may not always be worth the effort because of relatively modest analytical demands for the final standards. In that case, it may be feasible to utilize a stable and "inert" second component to monitor a less stable or reactive component or both. The use of such an internal standard requires a method of analysis wherein the sensitivity of one component relative to the other remains constant and each component can be detected separately.

Examples of the use of internal standards are:

1. vinyl chloride monomer (VCM) and propane (C_3H_8) in nitrogen (N_2),
2. acrolein (C_3H_4O) and C_3H_8 in H_2, and
3. methyl bromide (MeBr) and C_3H_8 in N_2.

In these instances, the behavior of binary mixtures of vinyl chloride in nitrogen, acrolein in nitrogen, and methyl bromide in nitrogen in gas cylinders was not known while the behavior of propane was well characterized [4,5]. Mixtures can be prepared to contain VCM and C_3H_8 in N_2, acrolein and C_3H_8 in N_2, and MeBr and C_3H_8 in N_2, where the concentration of the C_3H_8 is about the same as that of the other component of interest (either VCM or acrolein or MeBr). A gas chromatograph with a flame ionization detector (GC/FID) can be used to both separate and detect the two components of interest in these mixtures and the ratio, $R_{(I)}/R(C_3H_8)$, can be observed with time (where $R_{(I)}$ is response of the compound of interest and $R(C_3H_8)$ is the response of propane).

It should be emphasized that the method of analysis must not introduce an additional variable. For instance, if GC/FID is used to compare the substance of interest with a hydrocarbon internal standard, some means must be provided to assure that the sensitivity to both compounds remains the same. If not, the results may be quite misleading. Good laboratory practices in GC operation are sufficient to assure that the relative responses of the two compounds will remain constant. These include (a) using the same GC and conditions for the time study, (b) ensuring that the fuel gas flows to the flame are the same each time, and (c) ensuring that the detector is clean and working well.

An external standard may be also used. This procedure is quite similar to the one just described, but instead of having both the compound of interest (that is, VCM, MeBr, etc.) and the standard (that is, C_3H_8) incorporated in the same mixture, the experiment is performed with two cylinders. A cylinder mixture of the compound of interest will be prepared in nitrogen and a mixture of C_3H_8 in N_2 at about the same concentration in another cylinder will be used to study the stability of the mixture of interest. The same analytical procedures and precautions, as just mentioned, would then be followed.

Chemical Conversion

Another approach to the difficult-to-determine compound consists simply of chemically converting the substance to a more easily measured or more stable substance. Probably the best known applications of this technique are the catalytic conversion of carbon monoxide to methane with subsequent measurement of the methane by FID and the catalytic reduction of NO_2 to NO with subsequent measurement of NO by chemiluminescence.

Another example of conversion is the measurement of water vapor as the basis for the determination of trace quantities of either hydrogen or oxygen in compressed gases. The technique requires an absolute method for the measurement of water vapor. Formerly, the Weaver conductiometric method was used, but now coulometric instruments are employed. Before the Hirsch cell was developed for the measurement of oxygen, the detection of traces (0 to 10 ppm) was difficult. However, water (H_2O) at corresponding concentrations could be measured quantitatively. A system can be used that consists of a flowing sample gas stream, a means for adding hydrogen to the stream, a hot palladium catalyst, and finally a water vapor analyzer. A small amount of H_2 (less than 1%) is blended into the gas containing the oxygen to be measured as it flows slowly out of a cylinder and this is passed around the furnace to the detector. After equilibrium is reached, the gas is passed through the catalytic furnace where the added H_2 will react with O_2 to form H_2O and the increase in H_2O is observed. By substituting O_2 for the H_2, traces of the H_2 may be measured. This experiment works whether O_2 or H_2 or both are present at low levels in a gas.

Ozone (O_3) provides a different example of chemical conversion. This gas is very reactive and cannot be stored as a mixture in a cylinder. The standard is generated by exposing a flowing air stream to an ultraviolet lamp that provides enough energy to convert some of the O_2 to O_3. The O_3 produced exits the generation system and can be measured wet chemically by using the neutral buffered potassium iodide method or instrumentally by using an ultraviolet photometer. The concentration of O_3 exiting the generator can be varied by varying the air flow rate, by varying the power to the lamp, and by varying the amount of the lamp that is exposed to the air.

Conclusions

The usefulness of any one of these methods will depend on the particular application and, generally, the success will depend on the dedication, experience, and skill of the analyst. It cannot be overemphasized that in dealing with reactive or difficult gas mixtures, the two most important attributes of the analyst are experience and imagination. Seldom can two different gas mixtures be treated in the same way, but the experience gained from a study of one gas can be very helpful in studying another, especially in regard to adverse results.

In summary, a special calibration system for a reactive gas mixture is any system that will produce a mixture of the required concentration, of acceptable uncertainty, and of stability sufficient for the analysis at hand. The only way of being certain that a system performs as required is to devise a test procedure for the particular system.

References

[1] Hughes, E. E., Dorko, W. D., Scheide, E. P., Hall, L. C., Beilby, A. C., and Taylor, J. K., "Gas Generating Systems for the Evaluation of Gas Detecting Devices," NBSIR 73-292, National Bureau of Standards, Gaithersburg, MD, 1973.
[2] Nelson, G. O., *Controlled Test Atmospheres: Principles and Techniques*, Ann Arbor Science Publishers, Ann Arbor, MI, 1972.
[3] Hughes, E. E., Dorko, W. D., and Taylor, J. K., "System for Preparation of Known Concentrations of Methane in Air," NBSIR 73-255, National Bureau of Standards, Gaithersburg, MD, 1973.
[4] Scheide, E. P., Hughes, E. E., and Taylor, J. K., "A Gas Dilution System for Acrolein," NBSIR 73-258, National Bureau of Standards, Gaithersburg, MD, 1973.
[5] Scheide, E. P., Hughes, E. E., and Taylor, J. K., "A Gas Dilution System for Methyl Bromide," NBSIR 73-259, National Bureau of Standards, Gaithersburg, MD, 1973.

Workplace Atmospheres

Sam K. Norwood[1]

Strategy for Industrial Hygiene Monitoring in the Chemical Industry

REFERENCE: Norwood, S. K., **"Strategy for Industrial Hygiene Monitoring in the Chemical Industry,"** *Sampling and Calibration for Atmospheric Measurements, ASTM STP 957,* J. K. Taylor, Ed., American Society for Testing and Materials, Philadelphia, 1987, pp. 141–148.

ABSTRACT: The traditional approach to industrial hygiene sampling involves the subjective evaluation of factors in the work environment such as the likelihood of exposures during certain job tasks, the frequency of such tasks, the toxicity of the chemicals encountered, and the type of sampling needed to adequately quantify the particular exposures of interest. This approach to sampling requires judgment based on past experience and is most effective when there is adequate communication between industrial hygienists, physicians, toxicologists, analytical chemists, safety engineers, and process engineers. Recently, sampling strategies have incorporated statistical methods to aid in data interpretation and the selection of sample size. However, reliance on a statistical approach alone will not suffice to determine optimum sampling strategies because the objectives of industrial hygiene programs will usually extend beyond compliance to exposure guidelines. A sampling strategy should be founded on the traditional approach, but incorporate statistical methods where applicable. This balanced approach to sampling design offers a better utilization of industrial hygiene resources while accomplishing the objectives of the total industrial hygiene program.

KEY WORDS: industrial hygiene, sampling, air quality, calibration, atmospheric measurements

The techniques for evaluating exposures have changed in many ways over the years. Early industrial hygienists often encountered work situations that were obviously unhealthy. Many times actual air measurements were not needed to determine if there was a health problem because the employees were exhibiting symptoms of chemical toxicity. Furthermore, the existing state of the art of sampling and analytical techniques made it very difficult to make measurements and there was a significant reliance upon the judgment of the industrial hygienist.

In modern chemical plants, employee exposures are generally maintained well below levels that are recognized as producing adverse health effects and a sophisticated epidemiology study may be required to determine whether or not a health problem exists. Even so, employees at all levels are becoming increasingly more concerned about the chemicals used in a plant, the toxic properties of these chemicals, and the exposure levels associated with the jobs in a plant. Consequently, it is necessary for the industrial hygienist to document exposures in a comprehensive manner either by judgment or by actual measurement. Not only must potentially high exposures be documented, potentially low exposures must also be documented.

An evaluation of exposures in a plant may require a complex matrix of evaluation techniques. The types of processes, the types of exposures, and available sampling methods

[1] Industrial Hygiene Laboratory, The Dow Chemical Company, Midland, MI 48674.

impact the sampling strategies to be used. For example, whether the process is batch-operated or continuously operated will have an impact on the number and type of short-term excursion exposures encountered by a plant operator.

In the 1950s and before, chemical processes were batch processes requiring the opening of kettle covers, manually draining kettles, changing filters and flanges, operating centrifuges or wheels, shoveling wet cake after filtering, cleaning stills, and packaging without local ventilation. To properly conduct an industrial hygiene survey in those days, the industrial hygienist spent a considerable amount of time with the operators because the sampling and analytical techniques required the industrial hygienist to hold the monitoring equipment to obtain a sample in the breathing zone. Operators spent most of their time out in the process area and only a small fraction of their time in the control room or other nonproduction areas. From the 1960s through the 1970s, continuous processes came into use with improved stripping and recycling of volatiles. Automatic feeding systems, use of sight glasses, and other process changes occurred to reduce employee exposures. Operators working in these continuous processes spent much more of their time in the control room, with few operations requiring opening equipment. Today's processes are often computer-controlled or cubicle-controlled and operators may spend most of their time in the control room. However, there are specific operations and exposures that need to be evaluated during the small fraction of their time spent outside the control room. Time needs to be spent investigating the specific types of excursion exposures still inherent with chemical processing.

Improvements in sampling techniques and innovative instrumentation have led to changes in the practice of industrial hygiene. In the 1950s, the air sampling techniques generally used wet chemical analyses, requiring collection using glass absorbers and impingers. Limited sensitivity was obtained with titration or colorimetric analysis. In the 1960s, solid sorbents such as silica gel and activated carbon were starting to receive more attention as collection media. With the advent of gas chromatography, emission spectrometry for metals, X-ray diffraction for crystalline silica, and mass spectrometry, analytical capabilities were greatly expanded. In the late 1960s and early 1970s, portable direct-reading instruments, such as halide meters, mercury detectors, and portable infrared monitors began to be used routinely. Today, with the current advances in microprocessor technology and computerized data handling systems, the turnaround time from sampling to analysis to evaluation has increased at the rate where one can hardly stay abreast of developments.

The ability to collect and analyze data has increased tremendously. However, there has been a corresponding increase in the risk of collecting more data than are needed to make a statistical decision and selecting inappropriate sampling techniques for correctly evaluating a workplace. There is a tendency to collect a lot of data because it looks like a good industrial hygiene program when in reality there may be gaping holes in the data base. A systematic approach to comprehensive surveys, coupled with chemical-specific ongoing programs, ensures that the quantity and quality of the data base are reviewed periodically.

Initial Evaluation of the Workplace

One of the most important aspects of conducting a thorough comprehensive survey is taking the time at the very beginning to fully understand the process and to search out potential exposure conditions. This requires becoming acquainted with the work patterns of the operating personnel. At times, this can take several days, but it is time well spent.

A description of the process is a useful tool for planning a comprehensive survey. One should prepare flow sheets and floor plans, and study the handling of raw materials, intermediates, and final products.

For industrial hygiene purposes, the key elements in a flow sheet are the sources of potential exposure. The flow sheet should emphasize possible contact points, such as sample taps, vents, manholes, and drains—places where the operator can be exposed to the process chemicals. A line diagram should be made to show the progression of the raw materials through the intermediate steps to the final product. In most cases, there is no need to show the amounts of chemical being processed. However, in some cases, particularly in open systems, the concentration of the critical compounds should be indicated. Where a material leaves the system, the disposition should be indicated (another process, sludge sewer, etc.). The packaging operation should be briefly discussed (for example, manual loading of drums), along with an indication of the storage facilities (warehouse, tank farm, etc.). A well-prepared flow sheet will spotlight the major areas of concern and will minimize the chance of a potential hazard being overlooked. The flow sheet also serves as a reference point to locate process changes that may need checking in subsequent investigations.

The availability of floor plans to locate the process equipment is also very helpful in preparing for an industrial hygiene survey. Used in conjunction with the flow sheet, the floor plan can provide valuable insight into determining where the problem areas are likely to be located. The floor plan, or layout, should be as detailed as possible, showing pumping stations, trenches, and the location of process sampling taps, as well as major process equipment. The floor plan should also include any features unique to the process that could be a source of exposure (filter clean-out stations, air-conveying systems, etc.). Besides pointing out potential problem areas, the floor plan is also a useful tool during the survey to locate sites where measurements were made during the survey. This simplifies the explanation of where the samples were taken for reporting purposes and enables other investigators to duplicate the work in subsequent surveys.

The process chemistry must also be reviewed when planning a survey. Quite often the by-products or contaminants produced during a chemical reaction are more critical from an industrial hygiene aspect than the reactants or products. In describing the chemistry of the process, the reactions and possible side reactions should be defined for each major piece of equipment. A complete understanding of the process is essential in preparing for an industrial hygiene survey. The sampling and analytical techniques are selected on the basis of what is expected to be found. Knowing what airborne contaminants might be present can be just as important as knowing what concentrations are likely to be encountered.

It is necessary to study job assignments carefully and consider normal operations, abnormal situations, and occasional specific tasks. In today's world of personal monitoring, it is very easy to obtain employee exposure information for normal operations and to neglect obtaining exposure information on days when occasional specific tasks occur. To avoid missing the important specific tasks, efforts should be made at the beginning of a survey—in the planning steps—to take time to study all job assignments and the work patterns for employees over a period of several days or weeks. Unless the industrial hygienist is willing to spend time in the plant, staying with the operating personnel throughout all of the shift period, significant potential exposure tasks can be missed. Regardless of the personal monitoring information, time distributions for all job assignments of operating personnel should be obtained for normal operations, but more importantly, the occasional specific operation.

Sampling Techniques

Prior to all well-planned industrial hygiene surveys, sampling methods should be developed by an experienced analytical chemist. There are a number of questions to be answered to make sure that the measurements to be taken are both reliable and within validated limits. As a minimum, air sampling parameters must be established, recovery studies for the adsorbents conducted, and validated methods of analysis assured. In addition, the effectiveness of the sampling system should be checked by sampling a known concentration of the contaminant of interest.

The importance of establishing quality assurance techniques and following through with them cannot be overstressed. Once the sampling equipment and desired flow rates have been selected, the entire system (for example, sampling device, tubing, and pump) should be calibrated. An excellent check on the validity of the field data is to conduct field spike experiments by spiking sample collection media that are then used in side-by-side sampling with field samples [1,2]. By statistically testing the amount recovered from these field spikes, one can check the validity of the laboratory analyses and the validity of the sampling method for the work environment in the concentration range sampled.

Inhalation exposure via the air pathway is usually the most significant route of exposure for most chemicals and, subsequently, air monitoring is generally the most important sampling technique used by industrial hygienists. There are several approaches for collecting air samples to estimate exposure levels throughout a work day (that is, source, full-shift, excursion, and area sampling). These methods have the same purpose, to estimate the average exposure and define peak exposures during a typical work day. Those collection techniques that directly measure the air in the breathing zone of an employee are the best methods for estimating inhalation exposures. In addition, it is often necessary to use several sampling methods concurrently to verify exposure levels and to provide information that one method alone cannot supply.

Source samples generally are diagnostic in nature and are used to identify the presence of a material, to locate a source of emission, or, in the case of continuous air monitors, used to identify leaks. Obviously, these samples are generally inappropriate for use as exposure estimates.

Since the advent of solid sorbent sampling with miniature air pumps, full-shift personal samples have been commonly taken. These samples give a good picture of the average exposure throughout the sampling period. However, these samples do not give information on the exposure profiles throughout the shift. Short-term concentration peaks that might be of interest are not documented.

Excursion, task, or short-term samples provide data on exposures during specific tasks where a higher concentration may be created in the breathing zone than occurs in the general plant air. Frequently, these samples are higher than personal full-shift samples and reflect peak concentrations rather than the average exposure. It is very important to do specific task sampling, particularly where chemicals may produce acute health effects. Task concentrations are often highly variable and exposures from tasks that are routine but infrequent are difficult to interpret from random full-shift personal sampling. Some tasks that should typically be evaluated in chemical plants include loading and unloading operations, process sampling, packaging and drumming operations, and infrequent but routine tasks such as filter maintenance or changing catalyst.

Area samples give an idea of the general background concentrations in a plant. Such samples may be representative of exposure to individuals in that area who do not perform tasks involving the monitored chemical, but often will be much lower than the exposures of individuals in that area who perform tasks involving the monitored chemical.

A better understanding of the relationships between the various types of air sampling techniques can be had by considering a simple two-component exposure model. One component of exposure comes from breathing the general area concentrations in the plant as the employee walks from one area to another or spends time in areas doing work that does not produce localized excursions in concentrations. The other component of exposure comes from breathing concentrations of chemicals generated from doing specific tasks.

In order to measure these various components, it is sometimes a good idea to conduct simultaneous personal, task, and area sampling. Task samples provide a good estimate of the concentrations generated during short-term tasks and area samples provide a good estimate of general-area concentrations. Appropriately combining task samples with general-area concentrations should provide a good estimate of personal samples. Simultaneous sampling using these different techniques can give assurance that all of the exposure parameters have been considered. In addition, this type of sampling gives an idea of the sources of variability in routine full-shift personal sampling over time.

Obviously, any single type of air sample could be used inappropriately by someone who does not understand the significance of the data or its relationship to the biological response being evaluated. Generally, an industrial hygiene monitoring program consisting mainly of full-shift personal monitoring data is not enough. Too often, monitoring of specific operations is ignored. Area and excursion monitoring are basic approaches to diagnostic sampling. In addition to the area samples in each of the work areas where the employee spends time, short-term samples of tasks should be collected to define exposures during specific operations. These values should then be compared to full-shift personal sampling data.

Skin absorption is another route of exposure that can be of concern. In some cases, surface contamination may be a significant pathway for chemicals to reach the skin. This may be particularly true for chemicals that readily absorb through the skin in toxic amounts or absorb through the skin and bioaccumulate. Wipe tests may be taken to determine the levels of contamination on work surfaces. However, it must be remembered that the relationship between wipe tests and exposure is very complex. Therefore, the general use for wipe tests is as an index of cleanliness. The results are used to monitor the buildup of contamination so that surfaces can be cleaned before high levels of contamination have accumulated.

Biomonitoring techniques can be used in some situations to estimate exposures via the skin absorption route. However, a validated biomonitoring technique must be available. Not only must the analytical method be valid, there must also be a validated relationship between actual exposures in humans versus the biomonitoring results. Unfortunately, there are only a very few biomonitoring methods that are valid at exposure levels that might be encountered in an occupational setting.

Sampling Approach

The traditional approach to industrial hygiene sampling involves the subjective evaluation of factors in the work environment such as the likelihood of exposures during certain job tasks, the frequency of such tasks, the toxicity of the chemicals encountered, and the type of sampling needed to adequately quantify the particular exposures of interest. In using a traditional diagnostic sampling approach, one might most effectively use a combination of time weighted average (TWA) personal sampling, excursion sampling, area sampling, or source sampling to evaluate a particular situation. This approach to monitoring requires judgment based on past experience, and typically several years of apprenticeship under an experienced industrial hygienist is needed before an individual can make consistent judgments about work environments.

In the traditional sampling approach, a matrix of job/chemical combinations that require monitoring is developed from the initial evaluation of the workplace. However, sampling approaches should also incorporate statistical methods to aid in data interpretation and selection of sample size [3]. From a statistical perspective, a tolerance limit approach should probably be used to evaluate employee multiday exposures to chemicals with potential adverse health effects that are highly dependent on the rate at which the dose is delivered, for example, chlorine [4,5].

Additionally, a confidence limit approach may be advocated for evaluating exposures to chemicals with potential adverse health effects that are, for practical purposes, independent of the rate at which the dose is delivered when excursions are reasonably controlled, for example, lead. Clearly, judgments are involved in determining the appropriate statistical approach. In addition, the quantity and quality of toxicology, epidemiology, and medical data used in setting the exposure guidelines must be considered. However, relying on a statistical approach alone will not suffice to determining optimum sampling strategies because the objectives of most industrial hygiene progams must extend beyond compliance to exposure guidelines alone. For example, it is often cost-effective to take a few samples of the lower exposure groups when conducting surveys. The percentage difference in total costs is often insignificant in comparison to the strength that such data adds to professional judgment that a group is truly a low exposure group.

To investigate the statistical impact of potential sampling strategies within The Dow Chemical Company, models of some possible plant environments were developed and computer sampling simulations were run. Assuming a log-normal distribution for exposure data and a normal distribution for analytical data, standard deviations for those data were varied. In some of the simulation runs, a subpopulation of the 10% of the employees was entered whose average exposure was twice that of everyone else. When the simulation runs were divided into roughly two exposure profiles, typical and extremely high variability, some generalizations were made. For work environments that have typical variability, four to six samples were generally adequate for making decisions about a job/chemical combination requiring monitoring. This conclusion was based on both observation of the computer simulation results and statistical theory. However, the model appeared to breakdown when there were highly variable conditions including subpopulations with rare events.

This emphasized the need to do adequate task analysis. For example, suppose that the rate events that resulted in high exposures were filter changes that occurred once per month. If four random samples were taken throughout the year, there would have been about a 13% chance of sampling one or more days when filters were changed. If a sample were taken during a filter change, then one could have made the assumption that such exposures occurred 25% of the time when they really only occurred 3% of the time. Obviously, random sampling coupled with the statistical model did not work in this case, but sampling based on task analysis coupled with appropriately applied statistics would have worked.

Some actual plant sampling data helps illustrate some points about statistical sampling strategies. Table 1 contains TWA personal samples for methylene chloride. A reasonable industrial hygienist would conclude after reviewing these data that there probably was not an exposure problem. However, when these samples were statistically analyzed, the best estimate of the 95 percentile (mean plus the product of the Student's t-value and the standard deviation) was about 250 ppm. Obviously, the log-normal statistical model was not working properly. Notice that the third sample was almost eight times lower than the other samples, causing an estimated geometric standard deviation of 3.3. The model assumes that the data were skewed to the high side rather than the low side. This resulted in a high estimate of the 95th percentile. In effect, the model converted the 2 to 200 ppm. The statistical model

TABLE 1—*Plant data results for methylene chloride full-shift personal measurements (ppm).*
Exposure guideline is 100 ppm.

Individual Measurement, ppm	Arithmetic Average	Geometric Standard Deviation	Best Estimate of the 95th Percentile
15.3	. . .	. . .	. . .
14.6	15	1.03	18
1.88	11	3.31	247

TABLE 2—*Plant data results for methylene chloride full-shift personal measurements (ppm). Exposure guideline is 100 ppm.*

Number of Samples	Arithmetic Average	Geometric Standard Deviation	Best Estimate of the 95th Percentile
2	15	1.0	18
3	11	3.3	247
4	10	2.7	76
6	9	2.1	34
8	12	2.4	46
10	12	2.2	39
12	12	2.0	36

did not hold for this small set of samples. Generally speaking, one must always be careful when using statistics with small sample sizes. However, collecting 20 or 30 samples to be statistically sure would not be cost-effective when good judgment had the right answer much earlier. Again, a blend of judgment and statistics was needed.

Table 2 contains the same plant data that is in Table 1, except that more of the samples that were taken are shown. Notice that a good estimate of the work place environment has been made after collecting six samples and the statistical model produced results that fit past experience in this plant very well. Many years of data indicate an arithmetic mean of about 10 ppm with an occasional sample in the 30 to 40 ppm range. No samples have been collected above 40 ppm for this job.

A final step in analyzing the data is to evaluate the future monitoring needs in a plant. This analysis can be divided into two components. First, one must decide how many samples are needed to redescribe the exposure profile. This could be a statistical evaluation based on the past data or perhaps one could use a generalized rule of four to six samples for each job/chemical combination requiring monitoring. Once the sample size has been selected, the frequency at which this number of samples is taken must be determined. The sampling period may be yearly for chemicals near the exposure guideline. However, if the sampling results are far from the exposure guideline, one may never take another sample or only infrequent confirmation samples. Between these extremes, samples may be taken on a two-to-five year basis depending on the potential hazard associated with the chemical. The samples could be taken in one year during this two-to-five year period or sampling can be spread out over the entire sampling period. Experience has shown that a comprehensive survey about every three years, coupled with an ongoing survey for chemicals needing annual monitoring, is probably the best combination of sampling frequencies.

Again, judgment is needed and there are many modifiers to the statistical approach that should be considered. The toxic properties are important. If the exposure guideline is based on a bad smell with a large safety factor between the bad smell and organic damage, there is less of a need to sample as frequently as one would where there is a smaller safety factor between the exposure guideline and organic damage. Chemicals with good warning properties may need to be sampled on a less frequent basis than chemicals with no warning properties. Obviously, if there are major process changes, a new round of sampling may be needed. Less frequent sampling may be needed where there is a solid past data base as compared to a situation with only sketchy historical data. Finally, regulations, complaints, and new toxicology, medical, or epidemiology data may trigger additional sampling or reanalysis of the past data base.

Summary

First and foremost, industrial hygiene can have a favorable effect on the health and well-being of employees. Second, the comprehensive survey is the basic tool for evaluating exposures. A few guidelines for conducting such a survey follow. Develop job and process descriptions from an industrial hygiene perspective, making sure descriptions include all the chemicals. Identify where these chemicals appear in the process and how the tasks individuals perform relate to potential exposures to these chemicals. Document the sampling strategy used. In particular, document the justification for sampling or not sampling a job classification. It may seem obvious at the time, but will be difficult to reconstruct in the future when needed for an epidemiology study. Also, perform limited sampling of low exposure job classifications. Adding a few extra samples will not contribute significantly to preparation time, the time spent in the plant, or an analyst's time in analyzing a batch of samples. Take samples over time. Where concentration excursions are of concern, take samples throughout a shift. A good procedure is to take a full-shift sample in conjunction with short-term samples. Take samples over days, spreading samples across the seasons. Day-to-day variations are generally the largest source of variability in estimating exposures. Also, take samples over multiple years to document the subtle changes that may occur in a process.

The traditional and statistical approaches should be combined. Judgment is always needed, but statistics aid in making better decisions as long as one keeps in mind the assumptions being made and the underlying processes that result in exposures. Four to six representative samples spread throughout time are generally adequate for decision-making. For those job/chemical combinations that require periodic monitoring, this should usually be done on a one-to-five year basis, depending on the potential risk of each situation. However, a comprehensive survey every three years, coupled with an ongoing survey for chemicals needing annual monitoring, appears to be a practical goal.

References

[1] Chapman, L. M., Ward, B. G., and Jeannot, P. M., *Journal, American Industrial Hygiene Association,* Vol. 41, 1980, pp. 630–633.
[2] Borders, R. A., Melcer, R. G., and Gluck, S. J., *Journal, American Industrial Hygiene Association,* Vol. 45, 1984, pp. 299–305.
[3] Leidel, N. A., Busch, K. A., and Lynch, J. R., *Occupational Exposure Sampling Strategy Manual,* National Institute of Occupational Safety and Health, Cincinnati, 1977.
[4] Tuggle, R. M., *Journal, American Industrial Hygiene Association,* Vol. 43, 1982, pp. 338–346.
[5] Rappaport, S. M., Selvin, S., Spear, A. C., and Keil, C., *Journal, American Industrial Hygiene Association,* Vol. 42, 1981, pp. 831–838.

Richard G. Melcher[1]

Laboratory and Field Validation of Solid Sorbent Samplers

REFERENCE: Melcher, R. G., **"Laboratory and Field Validation of Solid Sorbent Samplers,"** *Sampling and Calibration for Atmospheric Measurements, ASTM STP 957*, J. K. Taylor, Ed., American Society for Testing and Materials, Philadelphia, 1987, 149–165.

ABSTRACT: Solid sorbents are being used extensively to sample contaminants in air. This technique has proven valuable, not only to the industrial hygienist because the solid sorbent sampling tubes are easy to use and transport, but also to the analytical chemist who finds the analytical procedures straightforward and adaptable to a wide range of compounds. Since the integrity of the sample depends on the nature of the compound and sorbent and on the effects of a number of other parameters, it is important, if not essential, to obtain laboratory validation data before using the technique for obtaining monitoring data. Field spiking and conformation studies are also recommended to determine the effects of the specific plant atmosphere and other sampling variables under actual field conditions. This paper gives guidelines for designing both laboratory and field validation studies.

KEY WORDS: air quality, calibration, sampling, atmospheric measurements, industrial hygiene sampling, solid sorbent sampling, charcoal tube air sampling, validation of air sampling methods, field validation of air sampling methods, solvent desorption

Solid sorbents are being used extensively to sample contaminants in air. A small tube containing a solid sorbent is convenient to use, can concentrate trace contaminants, and can be worn by a worker to determine breathing zone concentrations. Because solid sorbents are so convenient to use and transport, methods using solid sorbents are generally preferred over whole air and impinger methods for many compounds. There are two basic techniques for collection of substances in air using solid sorbents. The most widely used technique utilizes a small pump to draw the air sample through a bed of solid sorbent. The second technique, called passive or diffusional monitors, utilizes diffusion of compounds into a chamber containing a solid sorbent. The compounds can be recovered from the sorbents in both techniques by desorption with a suitable solvent or in some cases thermal desorption. Several reviews [1,2] describe broad application of these techniques. This paper will cover only the first technique, pump and tube samplers.

Before we get into a detailed description of the laboratory and field procedures, a broader picture of the method development/validation process as a whole will be discussed. Also, in order to get a better understanding of collection of compounds with solid sorbents, capacitive and volumetric breakthrough parameters that affect collection and parameters that in turn affect recovery will be discussed.

[1] Research Associate, Michigan Applied Science and Technology Laboratories, Analytical Laboratory, The Dow Chemical Company, Midland, MI 48640.

Progressive Method Development and Validation Steps

Table 1 lists the typical development steps for an air sampling method. In the first step, the project is defined, the needs of sensitivity and specificity are determined, and analytical and collection systems are suggested or extrapolated from the literature. In the second step, the initial experimental work tests the collection and analytical procedures and evaluates some of the important parameters. The third step is a systematic validation study after a tentative method is formulated. After the validation data are evaluated, some modifications of the method may be necessary.

Methods that are obtained from other laboratories or from the literature may include much of the required information. It is essential, however, to demonstrate the suitability of the method in your laboratory for a specific application, through prepared knowns and a field validation study.

The total monitoring method can be divided into the analytical procedure and the sampling procedure. If we define the analytical procedure as the determination of specific compounds in the desorption solvent, then either procedure can be changed independently as long as the interfacing parameters are considered, and the appropriate validation data for the revised section obtained. However, an adequate analytical procedure must be available before collection and recovery data can be obtained.

An analytical technique must be selected and tested before sampling when new methods are being developed. Table 2 lists analytical parameters that should be determined before and during the testing of new sampling methods.

Solid Sorbent Collection—Solvent Desorption

In selecting a sorbent for high collection efficiency, the recovery of the compound must also be considered. Often a mutual compromise may be necessary to obtain an acceptable

TABLE 1—*Progressive development and validation steps.*

Suggest Procedure (little or no experimental work)
1. Chemical, physical, and toxicological properties listed.
2. Concentration range and required sensitivity estimated.
3. Sorbent-desorption system(s) extrapolated from similar compounds.
4. Analytical procedure suggested.
5. Possible interferences indicated.

Tentative Procedure (limited experimental work)
1. Analytical procedure tested.
2. Desorption efficiency determined.
3. Optimized desorption time.
4. Preliminary breakthrough and bias study.
5. Preliminary field samples.
6. Effect of interferences, coadsorption, and temperature (if indicated).

Validated Method (systematic study)
1. Determination of breakthrough volume at concentrations of 2 times exposure guideline and relative humidity of 80% or greater.
2. Five samples of each concentration; 0.1, 0.5 (optional), 1, and 2 times exposure guideline for relative humidities less than 50% and greater than 80%.
3. Ten samples at 1 times exposure guideline stored for three to five weeks, five at room temperature, and five at refrigerator temperature.
4. Statistical evaluation of data.
5. Field validation by comparison to accepted method or by field spiking.
6. Collaborative studies with other laboratories.

TABLE 2—*Interfacing analytical and sampling procedures.*

Sensitivity	detection limit compatible with amount of compound to be collected
Specificity	interferences in the area and on a broader scale investigated
Reproducibility of analysis	over concentration range to be studied
Reproducibility of apparatus	operating conditions well defined and columns, reagents, etc. widely available
Adaptability	compatible with desorption solvents and other reagents
Simplicity	rapid, uncomplicated procedure
Automated	automatic injection and data handling systems

system. Table 3 lists some of the most used sorbents for the various types of compounds collected, although overlap allows some flexibility when sampling mixtures. Some sorbents, such as charcoal, have wide applicability, while some sorbents are designed for a specific type of problem. Tubes containing 150 and 600 mg of sorbent are commonly used for personal sampling; however, for long-term sampling of high concentrations or highly volatile compounds or both, tubes containing 1 g or more may be necessary.

Without getting too deeply into adsorption theory, two types of sorption processes may be considered when discussing collection of substances from air. Volumetric breakthrough usually occurs when sampling low concentrations on porous polymers, and, therefore, it will be discussed in more detail in the next section under thermal desorption. The phenomenon that usually controls breakthrough of most organic compounds when using activated carbon is capacitive breakthrough. The compounds are strongly held by carbon and saturation (capacitive breakthrough) occurs before volumetric breakthrough for most compounds. This type of capacitive breakthrough is also observed for other sorbents and depends on the sorbent and volatility of the compound [3,4].

TABLE 3—*General sorption-desorption systems for organic compounds.*

Sorbent	Desorption Solvent	Types of Compounds
Activated carbon	carbon disulfide, methylene chloride, ether (1% methanol or 5% isopropanol sometimes added)	miscellaneous volatile organics: methyl chloride, vinyl chloride, and other chlorinated aliphatics, aliphatic and aromatic solvents, acetates, ketones, alcohols, etc.
Silica gel	methanol, ethanol, diethyl ether, water	polar compounds: alcohols, phenols, chlorophenols, chlorobenzenes, aliphatic and aromatic amines, etc.
Activated alumina	water, diethyl ether, methanol	polar compounds: alcohols, glycols, ketones, aldehydes, etc.
Porous polymers	ether, hexane, carbon disulfide, alcohols	wide range of compounds: phenols, acidic and basic organics, multi-functional organics, etc.
Chemically bonded and other GC packings	ether, hexane, methanol	specialized: high boiling compounds—pesticides, herbicides, polynuclear aromatics, etc.

Figure 1 illustrates the distribution curve of a compound through the sorbent bed and the capacitive breakthrough curve of the compound in the effluent. At the start of collection, the compound is distributed through a section of the sorbent as shown by Curve 1. After continued collection, this section becomes "saturated"; that is, an equilibrium is established with incoming concentration (Ci) where the vapor sorption and desorption rates are equal. This is shown by Curve 2. The sorption front has moved through the front section of the collection tube (L 2/3) and is being collected on the backup section. If collection continues, the equilibrium zone lengthens, and the compound begins to break through the collection tube at Time$_B$, as shown by Curve 3. The breakthrough curve, Curve 4, is generated by monitoring concentrations of the compound in the effluent, and when all the sorbent is "saturated," the effluent concentration approaches the input concentration Ci.

Figure 1 deals with the collection of one compound. When two or more compounds are collected, the compound most strongly held may displace the other compounds down the length of sorbent bed. This effect is very pronounced for silica gel since atmospheric water vapor, which is strongly adsorbed on silica gel, will displace compounds (especially nonpolar compounds) and promote premature breakthrough. Other factors must also be considered when using solid sorbent sampling [5]. Deteriorative effects on collection efficiency will be detected, after the fact, by high concentrations on the backup section. However, if the factors are recognized before or during sampling, a greater safety factor can be applied to the sample volume that will reduce the number of invalid samples and the need to resample. The following factors influence collection efficiency.

Size of Collection Tube—In general, if the size (amount of sorbent) is doubled, then breakthrough volume (capacity) is doubled.

Flow Rate—The effect of flow rate varies with the sorbent. If the flow rate is too high, a nonequilibrium condition occurs due to poor vapor contact, and poor collection efficiency will result.

Concentration—Breakthrough occurs sooner for higher concentrations. For charcoal, the empirical Fruendlich isotherm appears to apply [6]. This equation takes the form

$$\log T_B = \log a + b \log C \tag{1}$$

A straight line results if the log of the breakthrough time, T_B, is plotted against the log of the concentration, C. The line will have Slope b and intercept of Log a.

Humidity—In general, an increase in humidity will result in a decrease in breakthrough volume. The magnitude of this effect depends on the properties of the sorbent and sorbate.

Temperature—An increase in temperature will result in a decrease in breakthrough volume. There is little specific information available for the various sorbents. A general guideline has been suggested for charcoal that for every 10°C rise in temperature, the breakthrough volume will be reduced by 1 to 10%.

Coadsorption—When two or more compounds are collected, the compound most strongly held may displace the other compounds, in order, down the length of sorbent bed.

Migration—A false indication of breakthrough may be caused by migration of the compounds collected on the front section to the backup section over an extended storage period.

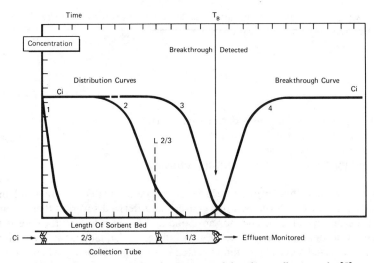

FIG. 1—*Concentration distribution in a solid sorbent collection tube* [5].

This can be reduced by refrigerating samples as soon as possible, or eliminated by using two tubes in series and separating them immediately after sampling.

Once the sample has been collected and returned to the laboratory, the collected compounds must be recovered from the sorbent for determination. Percent recovery is the percent of chemical recovered of the amount collected under actual sampling conditions. The major contributing factor is the desorption efficiency that is the partition at equilibrium of a chemical between a specific kind and volume of solvent and a specific batch and amount of solid sorbent. Recovery may also be influenced by other factors such as humidity, temperature, and compound stability to oxidation.

$$recovery = desorption\ efficiency \pm other\ factors \qquad (2)$$

Factors that cause changes in recovery are insidious, and unless these are understood and can be related to the chemical and physical properties of the compounds collected, errors can go undetected. The desorption efficiency is the most significant of the factors in defining the sorption-desorption system. Although desorption efficiency cannot always be isolated, it can be determined experimentally and is one of the first indicators of a potential problem in a suggested method. Other factors that may affect recovery are the temperature of the desorption solvent, moisture collected with the sample, coadsorbed compounds, desorption time, concentration of the collected compounds, and storage time [7–9].

For many systems, the desorption efficiency can be written in terms of an equilibrium constant; it is dependent on the ratio of solvent to sorbent for the distribution of the compound between the two phases. An equation has been derived by Dommer and Melcher [10] that relates the desorption efficiency to the volume of solvent and the amount of sorbent. The equation assumes the system is in equilibrium and can be approached from either direction. That is, the same desorption efficiency should be obtained when the compound is initially in the solvent or the solid phase. This has been shown to apply to most organic compounds in the concentration range of interest in Industrial Hygiene analyses. Desorption efficiency using the phase equilibrium method is similar to direct injection into the sorbent; only the test compound is prepared in the desorption solvent.

The following phase equilibrium equations can also be used to optimize the solid/liquid ratio when developing an analytical procedure.

$$\frac{1}{D} = K \frac{W_s}{W_l} + 1 \tag{3}$$

where

D = desorption efficiency (as a decimal fraction),
W_s = weight of solid phase,
W_l = weight of liquid phase, and
K = constant for specific compound and sorbent.

Once K is determined, the desorption efficiency can be calculated for any solid and liquid ratio. Posner [11] further developed this concept and derived several other equations for calculating the desorption efficiency for any solid and liquid ratio and for determining the ratio necessary to obtain a desired desorption efficiency. Equation 4 is a modification of Eq 3 and can be used to calculate the expected desorption efficiency when the volume is changed.

$$\frac{1}{Dn} = \left(\frac{1}{n}\frac{1}{D_1} - 1\right) + 1 \tag{4}$$

where

n = the ratio of solvent volumes n = new volume/initial volume,
D_1 = initial desorption efficiency, and
Dn = desorption efficiency using new volume.

Equation 5 can be used to select a volume to produce a desired desorption efficiency.

$$n_z = \left(\frac{1}{D_1} - 1\right) \frac{Z}{1 - Z} \tag{5}$$

Where n_z is the number of multiples of the original volume of desorbent needed to reach the desired efficiency, Z.

It should be noted that the partition ratio at equilibrium predicts the optimum desorption efficiency attainable, and other experiments may be necessary to detect nonequilibrium situations. Desorption efficiency should not be taken as the recovery since other factors may have a significant effect. After a solvent/sorbent system is selected and tested using the phase equilibrium technique, direct injections of the test compound are made into collection tubes with and without air being pulled through. If the desorption efficiencies as determined by direct injection are considerably lower than phase equilibrium values, interaction or reaction on the sorbent surface is indicated. If the total recovery from the simulated air collection is lower than the direct injection efficiency (even though no breakthrough has occurred), hydrolysis, oxidation, or another reaction may be indicated.

Since charcoal is such a good sorbent and is readily available, the solution to some sampling problems is to find a way to increase the recovery of the desired compound from charcoal. One way is by increasing the solvent/sorbent ratio as discussed in the phase equilibrium section. Two other approaches are the use of mixed solvents and the two-phase solvent system. In general, polar compounds usually show low recoveries from charcoal. By adding several percent of a polar solvent to the carbon disulfide desorbent solvent, recovery is often improved by 10 to 20%. Other approaches [12,13] use different mixed solvents such as 5% methanol in methylene chloride.

The mixed solvent technique has limited use for complex mixtures since it is more difficult to chromatograph, precludes determination of the polar solvent added, and may cause additional interference to other compounds present. A two-phase system has been developed [14] that is capable of measuring both polar and nonpolar organic solvents present simultaneously in work environments. The charcoal collection tubes are desorbed with a 50/50 mixture of carbon disulfide and water. After desorption, the water and carbon sulfide layers are analyzed separately. The high recoveries of the polar compounds are attributed to their partitioning into the aqueous phase after desorption from charcoal by carbon disulfide. Not only does the partitioning eliminate interferences of some polar and nonpolar combinations, but the partition coefficients give additional qualitative information.

Solid Sorbent Collection—Thermal Desorption

In the thermal desorption technique, samples are collected by pulling air through a tube containing a thermally stable sorbent bed. Instead of removing the sorbent and extracting with a solvent for analysis, the tube is heated, and the adsorbent compounds are purged directly into a gas chromatograph (GC). Thermal desorption eliminates use of solvents and other handling operations, is more sensitive than solvent desorption techniques, and the collection tubes are reusable. The main advantage of this technique is the high sensitivity obtained since the total sample collected in 1 to 3 L of air can be injected at one time. Sensitivity is in the low parts-per-billion range for most compounds.

A number of factors must be considered when selecting a sorbent for use with thermal desorption: (1) suitable collection properties, (2) thermal stability with repeated use, (3) low background contaminants during desorption, and (4) minimal decomposition or reaction during collection, storage, or analysis. A wide variety of sorbents have been evaluated including: activated charcoal and synthetic carbons; porous polymers such as Tenax, Chromosorb series, Porapak series, and XAD series; and liquid phase coated GC packings and bonded GC packings. Table 4 lists four sorbents and conditions that cover a wide range of volatility.

It is important to thoroughly condition each sampling slug before it is used by heating and purging with nitrogen. When determining trace quantities, the background pattern becomes quite important. Decomposition and oxidation products of the sorbent can interfere with the analysis, and extra care is needed in conditioning and in choosing desorption temperature and rate. Tubes are usually conditioned by heating in a flow of nitrogen (N_2) for 12 to 24 h [15–17].

The parameters that are often used in evaluating the collection properties of sorbents are the retention volume (peak maxima) and the breakthrough volume (first detectable loss). There are three techniques that can be used for determining the retention volume or breakthrough volume.

Temperature Extrapolation—The sorbent to be tested is packed into a collection tube and connected inside a gas chromatograph similar to a GC column. A compound is injected onto the column, and the GC retention volume is taken as the product of the flow rate and elution time (Fig. 2). Retention volumes are determined at several temperatures, and the log of the retention volume is plotted versus the reciprocal of the temperature ($1/K$). This plot is then extrapolated to determine the retention volume at ambient temperatures.

TABLE 4—*GC conditions for thermal desorption analysis.*

	Increase in Boiling Point of Collected Compound →			
Sampling tube	10 cm Carbosieve B adsorbent, 100/120	10 cm Porapak N porous polymer, 80/100	10 cm Tenax-GC porous polymer, 60/80	5 cm 20% DC-200 silicone oil on Chromosorb W HP support 100/120
Sampling tube desorption	5 min at 270°C	5 min at 200°C	5 min at 260°C	5 min at 230°C
Column	60 cm × 3 mm stainless steel, Carbosieve B adsorbent, 100/120	120 cm × 3 mm stainless steel, Porapak N porous polymer, 80/100	240 cm × 4 mm stainless steel, 10% OV-17 silicone oil on Gas Chrom Q 60/80	180 cm × 4 mm glass, 10% OV-17 silicone oil on Gas Chrom Q support, 100/120
Column temperature	5 min at 80°C program 20°C/min to 290°C and hold	5 min at 60°C program 15°C/min to 200°C and hold	5 min at 60°C program 15°C/min to 280°C and hold	5 min at 90°C program 15°C/min to 260°C and hold
Injection port temperature	80°C	120°C	220°C	200°C
Carrier gas flow, mL N$_2$/min	20	20	30	20

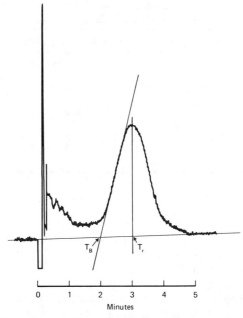

FIG. 2—*Graphic determination of breakthrough time* (T_B) *and retention time* (T_r) *using the gas chromatographic technique.*

Disappearances of Vapor During Purging—Collection tubes are loaded with a measured quantity of a compound and then purged with known volumes of air. The tubes are then desorbed, and the amount lost is determined by comparison to unpurged tubes.

Measurement of Breakthrough—The breakthrough of compounds under sampling conditions are monitored by analyzing the effluent from the tube directly using a GC detector or by attaching a back-up tube that is changed periodically and analyzed.

Capacitive breakthrough that is dependent on the total loading was discussed earlier. For low organic concentrations, as generally collected for thermal desorption, the breakthrough will occur as a function of total volume collected. This phenomenon, volumetric breakthrough, is related to GC retention volume (V_r). There appears to be a concentration (weight loading) for each compound on a specified sorbent at which the breakthrough changes from volumetric to capacitive [3]. Below this loading, the breakthrough volume is independent of the concentration and can be related to the retention volume of a single injected peak or in terms of the frontal elution of a continuous assault concentration. The sorbents used in thermal desorption are usually of lesser adsorptive strength than carbon and are used to collect much smaller amounts of chemicals. As a result, the temperature extrapolation is a useful tool for evaluating sorbents. The use of the temperature extrapolation technique has been described by a number of investigators [18]. The parameter usually calculated from the experiment is the specific volumetric capacity (V_g) in mL/g, which is calculated from the retention time, flow rate, and sorbent quantity.

$$V_g = \frac{F \times T_r}{g} \qquad (6)$$

TABLE 5—*Maximum safe sampling volume (MSSV) (tube size 10 cm × 6 mm (4 by 1/4 in.) stainless steel).*

	Tenax-GC, L/tube	Chromosorb 106, L/tube	Porapak N, L/tube
Methanol	0.003	0.15	0.4
Ethanol	0.04	0.8	4
Acrylonitrile	1	6	10
Acetone	1.6	4.2	8
Chlorobenzene	114	159	200
Toluene	21	97	117
Styrene	1190	455	645
Phenol	215	825	5900
Benzene	3.8	24	15

where

T_r = retention time in seconds to peak maxima,
F = carrier gas flow rate, mL/s, and
g = sorbent weight in grams.

After obtaining the retention volume at several temperatures, the log of retention volume is plotted versus $1/K$ and the curve extrapolated to 20°C or calculated by least-squares linear regression. The breakthrough volume, the point at which the compound is first detected, is dependent on the sensitivity of detection and the efficiency of the "column" (peak width), but the retention volume is independent of these factors. Taking into account the unknown effects of coadsorbed compounds under true atmospheric conditions, the maximum safety sampling volume (MSSV) can be estimated as being one-half the V_r. Table 5 lists the MSSV for a series of compounds and three types of sorbent tubes.

Simulated Atmosphere Test Systems

One of the main difficulties of developing and validating methods for monitoring low concentrations of chemicals in air is generating a simulated test atmosphere where variables such as concentration of the test compound, potential interfering compounds, and humidity can be accurately controlled. The closer the simulation is to the actual field conditions, the greater the probability that the validation procedure will be able to identify difficulties that may cause the method to fail in field application. Direct injection has been used for the development of methods for many compounds; however, this technique does not simulate actual sampling conditions. Several simplified techniques have been used to approach a dynamic system by injecting the compound into a U-tube, Fig. 3 [19]; glass wool, Fig. 4 [20]; or glass bulbs [21] prior to the sampling tube. For some compounds of low volatility, one of these techniques or direct injection may be the only choice. For stable, volatile compounds, a simulated atmosphere may be approached using plastic film bags. This approach is simple and suitable for many compounds, but conditions such as humidity are difficult to control, and a bag storage study is needed to determine suitability [19].

Diffusion tubes, permeation tubes, syringe pumps, vapor saturation, and special generation techniques have all been used. The diffusion tube is useful for compounds of medium volatility and are prepared from precision bore capillary tubing (~0.2 to 3 mm) and lengths of approximately 3 to 5 cm. Permeation devices are widely used for volatile compounds when low air concentrations are needed. Permeation devices control the release of chemicals

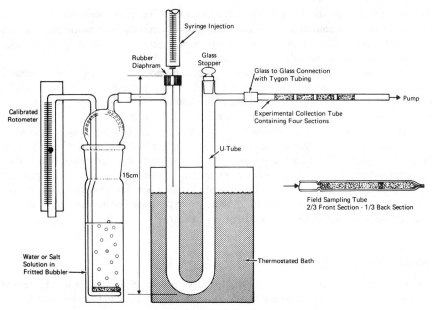

FIG. 3—*Dynamic U-tube technique for spiking collection tubes* [19].

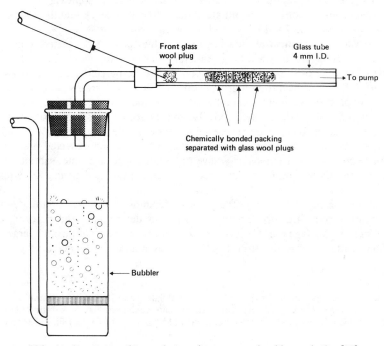

FIG. 4—*Dynamic spiking technique for compounds of low volatility* [20].

by permeation through the walls of a plastic membrane. The most commonly used materials for construction of permeation tubes are FEP (fluorinated ethylene propylene copolymer) TFE (tetrafluoroethylene polymer) because they are generally chemically inert, although other materials may be suitable for some chemicals [22,23].

A dynamic simulated atmosphere generation system more closely approaches actual field conditions, and a number of designs and approaches are reported in the literature [24]. The dynamic system usually consists of a number of components: (1) a filter system to produce purified air, (2) a flow/dilution system to control and blend the air streams, (3) a metering or generation system to produce known amounts of a chemical into the air stream, (4) a humidity generator and control system, and (5) a mixing and sampling chamber. Purified air is usually obtained by filtering pressurized air through molecular sieves or other drying adsorbents and charcoal beds to remove organic compounds. The system can be quite simple or more complex [25] depending on the degree of purity needed. The air flow is controlled using valves and rotometers, critical orifices, or mass flow controllers. Several systems containing these components are commerically available.

Laboratory Validation (Systematic Study)

Before starting the validation study, the tentative procedure (Table 1) should be worked out in detail. The preliminary testing should indicate a high potential for a successful method. Once a tentative procedure has been selected and a means for preparing accurate known samples has been devised, an in-depth validation study can be designed. A recommended design is given in Table 6. Various modifications may be necessary in the design if chemicals are reactive or hydrolyze easily. The minimum number of samples recommended for validation include the concentrations $1/10 \times$ PEL,[2] $1 \times$ PEL, and $2 \times$ PEL. If additional data are necessary, the concentration of $1/2 \times$ PEL is recommended. In some cases, the analytical method is not sensitive enough to quantitate at the $1/10 \times$ PEL level, and the concentrations $1/2 \times$ PEL, $1 \times$ PEL, and $2 \times$ PEL should be tested.

There are various statistical approaches to treat the data. A suggested approach and equations for treating the data will be given later. The first step would be to determine if a bias exists for some of the parameters by comparing the percent recovery and standard deviation for different concentration groups, different humidity groups, and for stored and non-stored samples. If all the data generally falls within the recommendations given in Table 6, a total treatment of the data may be possible to produce an overall method recovery and total relative precision (RP_T) value. These values, for example, recovery 87% ± 18% (95% confidence level), would be used to calculate the true value and expected precision for a sample. In some cases, restrictions may have to be put on storage time; sample volume in high humidity situations; and for variable recovery, an adjustable recovery value related to concentration used.

The total relative precision of an actual sample depends on the precision in determining a single measurement (Eq 10) and the precision in determining the percent recovery (Eq 11). When the average percent recovery, $\overline{Rn}$, has been calculated from a large number of data points, the contribution from Eq 11 has only a minor influence on the combined precision.

_____

[2] PEL (permissible exposure limits) is the Occupational Safety and Health Administration's (OSHA) time weighted average (TWA) guideline. Other appropriate guidelines may be substituted as desired, that is, American Conference of Governmental Industrial Hygienists' (ACGIH) TLV, European MAK, etc.

TABLE 6—Validation of sorbent tube/pump methods.

Parameter	Recommended Validation Experiments	Recommended Statistical Criteria
Desorption efficiency and optimum desorption time (by phase equilibrium or static direct injection)	$n = 10$; Concentration: 0.1 (or 0.5), 1, and 2× PEL; Desorption times: 15, 30, and 240 min	desorption efficiency at least 75% and independent of concentration. Bias between desorption efficiency and total recovery less than ±10% relative
Breakthrough	Check for breakthrough at 2× PEL; ≥80% relative humidity; (up to 8 h using recommended parameters)	breakthrough into back section less than 10% of total using recommended sampling procedure
Accuracy (% recovery) and total relative precision (for TWA exposures) (by dynamic direct injection)	$n \geq 15$; Concentration: 0.5, 1, and 2× PEL; Relative humidity: <50% (~75°F)	accuracy (% recovery) between 75 and 120%; total relative precision <±20% at the 95% confidence level
Excursions (% recovery and precision) necessary only if a different flow rate is used	$n \geq 15$; Concentration: 0.1 (or 0.5), 1, and 2× PEL STEL[a]; Relative humidity: <50% (~75°F)	accuracy (% recovery) between 75 and 120%; precision: <±20% at the 95% confidence level
Humidity effects	$n \geq 15$; Concentration: 0.1 (or 0.5), 1, 2× PEL; Relative humidity: ≥80% (~75°F)	bias due to humidity should be <±10% relative
Storage effects	$n = 10$, stored three to five weeks; Concentration: 1× PEL; five stored at refrigerator temperature; five stored at room temperature	bias due to storage should be <±10% relative (at least two weeks). Check for migration to back sections

[a] STEL is short-term exposure limit (ACGIH, OSHA, or European guides are acceptable).

Calculation of Recovery and Precision

$$R = \text{percent recovery} = \frac{\text{amound found}}{\text{amount added}} \times 100\% \tag{7}$$

$$\overline{Rn} = \text{average percent recovery for } n \text{ number of data points} \tag{8}$$

$$= \frac{\Sigma Rn}{n}$$

$$\sigma = \text{standard deviation} = \sqrt{\frac{\Sigma[(Rn - \overline{Rn})^2]}{n - 1}} \tag{9}$$

$$RPs = \text{relative precision of a single determination at the 95\% confidence level} \tag{10}$$

$$= \frac{t\sigma}{\overline{Rn}} \times 100\%$$

Where t is taken from statistical t-score tables for 95% confidence level (CL). If 20 or more datapoints are used, t approaches the value of 2.

$$RPn = \text{Relative precision of the calculated average percent} \tag{11}$$

$$\text{recovery, } Rn, \text{ at the 95\% confidence level}$$

$$= \frac{t\sigma}{\overline{Rn}\sqrt{n}} \times 100\%$$

$$RP_T = \text{total relative precision at 95\% confidence level} \tag{12}$$

$$= \sqrt{(RPs)^2 + (RPn)^2 + RP_p)^2}$$

Where RP_p = relative precision of sampling pump (usually estimated to be $\pm 5\%$ of the measured sample volume).

The precision of the sampling pump, RP_p, must also be included in Eq 12 if a mass balance type of recovery experiment (direct injection, weighed permeation tube, U-tube, etc.) was used instead of sampling a measured volume of a known concentration (plastic film bag, chamber, etc.) with a sampling pump. Often, RP_p is taken as $\pm 5\%$ when estimating RP_t.

Field Validation

The field validation discussed here is really a field confirmation, and it depends on having good laboratory validation data. Field validation is necessary to detect problems that are specific to the actual work site, since it is not feasible to study all the environmental variables in the laboratory. For field validation, it is necessary to prepare accurately spiked sampling tubes, which can be sampled side-by-side with a normal sample, to determine the recovery and whether the recovery has been affected. A recent publication [24] has given several approaches for field spiking experiments. If the amount of an added compound is recovered as expected from the laboratory data, a greater confidence in the suitability of the method is gained. Field spiking is not intended to detect chromatographic interferences that will be observed (but not necessarily recognized) during analysis, but to detect chemical or environmental interferences that may affect the collection, stability, or recovery of the analyte when monitored under actual field conditions. Field spiking also is a test of the integrity of the sampling, transportation, and analytical procedures. While field spiking may not reveal the specific causes of method problems, it will indicate when a problem exists.

The field spiking experiment is used to determine if the amount added is recovered as predicted by the laboratory experiments. This is done by sampling "side-by-side" one or more field samples and spiked field samples. After subtracting the amount found in the field sample, C_f, from the amount found in the spiked field sample, C_s, the remainder is compared to the amount added, C_c.

$$\% \text{ spike recovery} = \frac{C_s - C_f}{C_c} \times 100 \qquad (13)$$

The spike recovery will determine the total recovery of the compound even when the actual recovery of the amount collected in the field is unknown due to an effect of the workplace atmosphere. The external recovery expected is the same recovery found in laboratory experiments, $\%\overline{Rn}$ (Eq 8).

Before a decision can be made whether or not the spike recovery is acceptable, information about the precision of the experiment is necessary. Since a precision factor is involved in the collection and analysis of both the field sample and the spiked field sample, the significance in the recovery of the spike depends on the combined error from both samples. In order to determine if the spike recovery indicates a problem, the range of recovery expected from combining the two errors must be determined.

For a laboratory validated method, the total relative precision of the method, RP_T, can be calculated and includes all sources of variation for the method in the laboratory (Eq 12). The combined error, CE, can be defined as

$$CE = \sqrt{(C_s \times RP_T)^2 + (C_f \times RP_T)^2} \qquad (14)$$

If the RP_T is calculated for a given confidence level (that is, 95% CL), then the potential combined error is defined for the same confidence level.

The expected recovery limits, $\pm\%RL$, from which the validated recovery, $\%\overline{Rn}$, may vary based on the combined error is

$$\pm\%RL = \frac{\pm CE}{C_c} \times 100 \qquad (15)$$

In order to be valid at the stated confidence level, the recovery calculated from the experiment must fall within the expected recovery range (ERR):

$$ERR = \%R_v \pm \%RL \qquad (16)$$

One factor that influences the statistical significance of the field recovery determined is the number of field spikes and field samples taken in one experiment. Equation 15 is statistically correct for one field spike versus one field sample. A more general equation for more than a (1×1) experiment is

$$\pm \%RL = \frac{\pm CE}{C_c \sqrt{df}} \times 100 \tag{17}$$

The factor df is the square root of the degrees of freedom and depends on the number of field samples (n_f) and the number of field spikes (n_s).

$$\sqrt{df} = \sqrt{(n_f - 1) + (n_s - 1)} \tag{18}$$

For two field samples versus two field spikes (2×2), the recovery limit would be narrowed by $1/\sqrt{2}$ and for three field samples and three field spikes (3×3), by $1/\sqrt{4}$. This narrower (more precise) recovery limit would increase the significance in any difference observed in field spiking recoveries. A minimum of a 2×2 field spiking experiment is recommended.

A second factor that influences the statistical significance is the ratio of the amount added, C_c, to the amount in the field, C_f. If C_c is small compared to C_f, the relative error in determining the amount of C_c recovered from the field spike can be very large, since a large number, C_f, is subtracted from another larger number, C_s, to determine the relatively small number. The optimum situation is when $C_c \sim C_f$. Since this is difficult to accomplish in most field sampling situations, it is recommended that C_c be at least one half of C_f and not larger than $3C_f$.

Summary

When attempting to discuss method development and validation in general terms, it becomes quite complex because of the number of variables and variations one must deal with. It is important that the industrial hygienist, analytical chemist, and other working on the project communicate their expert information with one another so that the design for the total project can be optimized. At that point, Table 1 can be used as a guide to a successful method of development and validation.

References

[1] Melcher, R. G., *Analytical Chemistry*, Vol. 55, 1983, pp. 40R–56R.

[2] Melcher, R. G. and Langhorst, M. A., *Analytical Chemistry*, Vol. 57, 1985, pp. 238R–254R.

[3] Adams, J., Menzies, K., and Levins, P., "Selection and Evaluation of Sorbent Resins for the Collection of Organic Compounds," EPA-600/7-77-044, National Technical Information Service, PB-268 559, 1977.

[4] Certoni, G., Bruner, F., Liberti, A., and Perrino, C., *Journal of Chromatography*, Vol. 203, 1981, pp. 263–270.

[5] Melcher, R. G., Langner, R. R., and Kagel, R. O., *Journal*, American Industrial Hygiene Association, Vol. 39, 1978, pp. 349–361.

[6] Nelson, G. O. and Harder, C. A., *Journal*, American Industrial Hygiene Association, Vol. 37, 1976, pp. 205–215.

[7] Mueller, F. X. and Miller, J. A., *Journal*, American Industrial Hygiene Association, Vol. 40, 1979, pp. 380–386.

[8] Gagnon, Y. T. and Posner, J. C., *Journal*, American Industrial Hygiene Association, Vol. 40, 1979, pp. 923–925.

[9] Krajewski, J., Gromiec, J., and Dobecki, M., *Journal*, American Industrial Hygiene Association, Vol. 41, 1980, pp. 531–534.

[10] Dommer, R. A. and Melcher, R. G., *Journal*, American Industrial Hygiene Association, Vol. 39, 1978, pp. 240–246.

[11] Posner, J. C., *Journal*, American Industrial Hygiene Association, Vol. 41, 1980, pp. 63–66.

[*12*] Posner, J. C. and Okenfuss, J. R., *Journal*, American Industrial Hygiene Association, Vol. 42, 1981, pp. 643–646.
[*13*] Posner, J. C., *Journal*, American Industrial Hygiene Association, Vol. 42, 1981, pp. 647–652.
[*14*] Langvardt, P. W. and Melcher, R. G., *Journal*, American Industrial Hygiene Association, Vol. 40, 1979, pp. 1006–1012.
[*15*] Sydor, R. and Pietrzyk, D., *Analytical Chemistry*, Vol. 50, 1978, pp. 1842–1847.
[*16*] Dietrich, M. W., Chapman, L. M., and Mieure, J. P., *Journal*, American Industrial Hygiene Association, Vol. 39, 1978, pp. 385–391.
[*17*] Russell, J. W., *Environmental Science and Technology*, Vol. 9, 1975, pp. 1175–1180.
[*18*] Senum, G. I., *Environmental Science and Technology*, Vol. 15, 1981, pp. 1073–1075.
[*19*] Severs, L. W., Melcher, R. G., and Kocsis, M. J., *Journal*, American Industrial Hygiene Association, Vol. 39, 1978, pp. 321–326.
[*20*] Melcher, R. G., Garner, W. L., Severs, L. W., and Vaccaro, J. R., *Analytical Chemistry*, Vol. 50, 1978, pp. 251–255.
[*21*] Chapmann, L. M., Ward, B. G., and Jeannot, P. M., *Journal*, American Industrial Hygiene Association, Vol. 41, 1980, pp. 630–633.
[*22*] Waack, R., Alex, N. H., Frisch, H. L., Stannett, V., and Szwarc, M., *Industrial Engineering and Chemistry*, Vol. 47, 1955, pp. 2524–2527.
[*23*] Dharmarajan, V. and Rando, R. J., *Journal*, American Industrial Hygiene Association, Vol. 40, 1979, pp. 870–876.
[*24*] Borders, R. A. and Melcher, R. G., *Journal*, American Industrial Hygiene Association, Vol. 45, 1984, pp. 299–305.
[*25*] Barsocchi, A. T. and Knobel, R., *American Laboratory*, Vol. 2, No. 2, 1980, pp. 81–90.

Frederic Belkin[1] and Richard W. Bishop[1]

The U. S. Army's New Industrial Hygiene Sampling Guide

REFERENCE: Belkin, F. and Bishop, R. W., "**The U. S. Army's New Industrial Hygiene Sampling Guide**," *Sampling and Calibration for Atmospheric Measurements, ASTM STP 957*, J. K. Taylor, Ed., American Society for Testing and Materials, Philadelphia, 1987, pp. 166–175.

ABSTRACT: A new Industrial Hygiene (IH) Sampling Guide has been developed at the U. S. Army Environmental Hygiene Agency (USAEHA). Sampling procedures for over 100 compounds are described. The Guide contains minimum and maximum air volumes, recommended sampling rates and times for time weighted average (TWA), ceiling and short-term exposures, and the proper collection media for air sampling. It also addresses the types of containers and amount of sample required for bulk samples. This IH Sampling Guide is used by all Army industrial hygienists and industrial hygiene technicians involved in workplace monitoring. It enables the field personnel to sample using methods that are compatible with the Agency's state-of-the-art laboratory procedures.

KEY WORDS: sampling, industrial hygiene, air sampling, bulk sampling, toxic materials, air quality, calibration, atmospheric measurements

A new Industrial Hygiene (IH) Sampling Guide has been developed at the U. S. Army Environmental Hygiene Agency (USAEHA). Sampling procedures for over 100 compounds are described. The Guide contains minimum and maximum air volumes, recommended sampling rates and times for time weighted average (TWA), ceiling and short-term exposures, and the proper collection media for air sampling. It also addresses the types of containers and amount of sample required for bulk samples. This IH Sampling Guide is used by all Army industrial hygienists and industrial hygiene technicians involved in workplace monitoring. It enables the field personnel to sample using methods that are compatible with the Agency's state-of-the-art laboratory procedures.

The Army's IH Sampling Guide assures that uniform technical advice for routine industrial hygiene sampling is being dispensed to the industrial hygienists and technicians located at Army activities throughout the world. With this Guide, the field personnel have a consolidated list of all of the Army's commonly used sample collection procedures. As a result, the number of consultations on routine sampling procedures with USAEHA professional staff are reduced. Further, the Sampling Guide leads to a reduction in the number of samples being rejected by the chemistry laboratory because they are improperly sampled. The field personnel have written guidance on proper sampling and the laboratory managers have a written basis for rejecting incorrect samples.

A summary of the air sampling procedures for TWA and ceiling or short-term peak exposures are provided in Table A-1 of the IH Sampling Guide. The first page of this table

---

[1] Chemists, U. S. Army Environmental Hygiene Agency, Aberdeen Proving Ground, MD 21010-5422.

is illustrated here as Table 1. Table A-1 from the Guide also details the range of sampling rates and air volumes to be used for adsorbent tubes, impingers, and filters for numerous compounds. These give the sampler flexibility while providing the analyst sufficient amounts of sample for analysis. Maximum permissible collection volumes and flow rates are provided to decrease the probability of exceeding the capacity of the collection media, which would result in sample loss.

Guidelines for Collecting Air Samples

In addition to the sampling summary in Table A-1, there are general sections of the Guide devoted to description of proper sampling techniques using filters, adsorption tubes, impingers, and passive monitors. These sections address the usage or problems in sampling with each medium.

One common problem in filter sampling, for example, is the failure of the sampler to collect the minimum number of litres required for analysis, particularly for metals, dust, and oil mist. The USAEHA recommends the use of pumps that can sample up to 4 L/min. It is encouraged that the maximum sampling rate consistent with good pump operation be used to assure the meeting of the minimum recommended volumes. Many combinations of metals, such as lead and chromium, can be sampled and analyzed on the same filter. A list of metals requiring individual filters is provided in the Guide.

Another common problem that the Guide addresses is the collection of too high an air volume on adsorption tubes. This is particularly true of charcoal tubes that generally require small air volumes. Oversampling can cause breakthrough with the possibility of sample loss. High humidity, greater than 50% relative humidity, combined with high ambient temperatures, greater than ~30°C (85°F), or very high humidity, greater than 80% relative humidity, with normal ambient temperatures increases the problem of breakthrough. To reduce this probability of breakthrough and sample loss under the preceding conditions, one-half of the recommended maximum sample volumes (Table 1) should not be exceeded. This is a general rule; some polar compounds may have the maximum tube loading reduced by more than half under humid conditions. The 400/200 mg charcoal tube is used as the standard for USAEHA's industrial hygiene charcoal tube sampling. The use of these larger charcoal tubes reduces the frequency of changing tubes for full-term sampling. The 100/50 mg charcoal tube can also be used by collecting no more than one half of the maximum air volume and flow rate recommended for the larger tubes.

The USAEHA has sanctioned the use of passive monitors as an alternative method for selective charcoal tube procedures where manufacturers claims for passive monitors have been validated.

The USAEHA, itself, has validated four compounds in comparative studies of passive monitors with charcoal tubes [1,2]. Monitors are recommended for collection of the anesthetic gases halothane and enflurane because of the desirability to eliminate the industrial hygiene sampling pumps in hospital operating rooms [3]. Passive monitor's usage for collecting methyl chloroform and trichloroethylene has also been validated under laboratory and field conditions at degreasing operations.

Another section of the Guide is devoted to discussion of quartz (crystalline silica) sampling. It is recommended that only respirable silica be collected. This is because the respirable fraction is the pathogenic species (silicosis causing) and the calibration curve used in X-ray diffraction analysis is based upon respirable silica standards. The threshold limit value (TLV) for total quartz assumes that usually less than one third of the total quartz is respirable [4]. This assumption may not be correct. The polyvinyl chloride (PVC) membrane filters used

TABLE 1—Air sampling procedures for chemical contaminants.

Chemical Contaminant	Sampling Method	Sampling Rate or Time	Sample Volume in Litres		AEHA Procedure Number
			Minimum	Maximum	
Acetic acid	Chromosorb®P tube for acids (ORBO 70 acid tube)	100–500 mL/min	15	60	(1)
Acetone	200/400 mg charcoal tube	20–100 mL/min	1	6	(2)
Acid mists	see specific acid				
Alkali mists (such as NaOH, KOH)	see specific compound				
Aluminum	filter cassette, closed-face (CE 0.8 μm filter)	1–3 L/min	100	400	(3)
Ammonia	ammonia tube (ORBO 77)	100–500 mL/min	3	24	(4)
Amyl acetate	200/400 mg charcoal tube	20–500 mL/min	5	40	(5)
Aniline	260/520 mg silica gel tube	200–500 mL/min	25	60	(6)
Antimony compounds	filter cassette, closed-face (CE 0.8 μm filter)	1–2 L/min	100	1000	(7)
Arsine	200/400 mg charcoal tube	0.01–0.02 L/min	1	10	(8)
Asbestos	25 mm filter cassette, open-face shrouded with 50 mm extension cowl (CE 0.8 μm filter). See paragraph 2a	0.5–2.5 L/min	360	2000	(9)
	Note: For monitoring asbestos following abatement actions, sample at 1 to 5 L/min for a sample volume between 1300 and 3000 litres.				
Azide	see Hydrazoic acid				
Barium compounds	filter cassette, closed-face (CE 0.8 μm filter)	1–2 L/min	300	1000	(10)
Benzene	200/400 mg charcoal tube	50–500 mL/min	25	40	(11)
Beryllium	filter cassette, closed-face with spacer (CE 0.8 μm filter)	1–4 L/min	250	1000	(12)

for sampling are preweighed, so the percentage of quartz in the air may be determined from the ratio of the weight of quartz collected to the total weight gain of the filter.

Guidelines for Collecting Bulk Samples

Table A-2 in the Sampling Guide, illustrated here as Table 2, lists the procedures for collecting bulk samples. The compatibility of materials, storage, stability, and sample size requirements were considered when this list was established. Submission of bulk samples for asbestos has been a source of concern since a number of submissions were received in paper envelopes, many of which leaked. The USAEHA recommends that the samples be collected in screw cap glass vials to prevent contamination of our laboratory and exposure of our analysts to asbestos. A 12.7 by 12.7 mm (½ by ½ in.) section of the suspected bulk asbestos is recommended.

Field Blanks

The section on general sampling instructions of the Guide contains a discussion on field blanks, including their purpose and proper usage. For example, the guide recommends the use of a field blank with each set of air samples or one per ten samples. The USAEHA will not accept samples for analysis unless blanks are submitted. They should be carried to the workplace but should not be contaminated by intentionally exposing them to the workplace environment being sampled. The blanks should be from the same lot of material and prepared at the same time as the collection media, and those for adsorption tubes must have both ends broken and capped. The use of blanks enables the analyst to determine if there is significant contamination of the samples or interferences present in the sampling media. Interferences can be caused by problems with the reagent, filters, or tubes. The USAEHA has encountered a formate interference in one lot of commercial formaldehyde tubes. This interference could have led to false positive results if blanks had not been submitted. Contamination can occur during both storage and transportation. For example, contamination of charcoal tubes by storage of the tubes in a refrigerator with organic solvents has been detected by use of a blank tube. Gravimetric analysis for preweighed membrane filters are affected by changes in temperature and humidity [5]. The blank membrane filter enables the analyst to "correct" for these changes. It is requested that three blank filters be submitted with each set of filters for sodium hydroxide analysis since some lots of membrane filters have significant background levels of sodium [6]. These high background sodium levels will give false positive results because the samples are analyzed for sodium hydroxide by atomic absorption analysis for sodium. Not all samples can or should be corrected for a blank value. Blank correction may only be used if the blank is uniform on all samples and the blank value is small relative to the level of interest. In cases of large blank values or "unevenly" distributed contamination, the blank serves as an indication that the set of samples is invalid.

USAEHA Procedures in Guide

The IH Sampling Guide provides the best state-of-the-art sampling procedures compatible with laboratory methodology. Documentation has been provided in Table A-3 of the Guide, illustrated here as Table 3. Many of these procedures were developed at USAEHA and are not yet found in any other consolidated publication such as the National Institute for Occupational Safety and Health (NIOSH) *Manual of Analytical Methods* [7,8]. The Guide provides the necessary information to industrial hygienists on how to sample using these procedures. For example, USAEHA recommends collecting hydrochloric, hydrazoic, and

TABLE 2—*Bulk sampling procedures for chemical contaminants.*

Chemical Contaminant	Container Requirements	Sample Size
Asbestos	screw cap; glass or plastic vial; plastic bags are not acceptable	$\frac{1}{2}$ in. $\times$ $\frac{1}{2}$ in.[a] section
Corrosive (acidic or basic)	all glass (for acids only) or polyethylene (for acids and bases)	100 mL (unused material preferred)
Lead or chromium in paint	screw cap; plastic container	20 to 50 mL
Lead in paint chips	screw cap; glass or plastic container, or plastic envelope	1 g (a dime weighs about 2 g); do not submit plaster or other backing materials
Organic solvents including paints	all glass container, or glass container with Teflon-lined screw cap, or all metal can; do not use plastic or paper lined caps	100 mL (unused material preferred)
Pentachlorophenol in wood	wrap in aluminum foil	2 in. $\times$ 2 in. sections; do submit sawdust
Polychlorinated biphenyls (PCBs)	glass container with Teflon-lined screw cap	1 to 2 mL

[a] 1 in. = 25.4 mm.

TABLE 3—*Documentation of IH air sampling procedures.*

Chemical Contaminant	Sampling Method	Analytical Method	Coefficient of Variation	Reference	AEHA Procedure Number
Acetic acid	ORBO-70 tube	Ion chromatography	7.0%	AIHA J. 42(6):476–8 (1981)	(1)
Acetone	Charcoal tube	Gas chromatography, FID	8.2%	NIOSH: 1300 (3rd Ed.)	(2)
Aluminum	CE filter	Atomic absorption, flame	5.8%	NIOSH: 7013 (3rd Ed.)	(3)
Ammonia	Ammonia tube (ORBO-77)	Ion chromatography	8.4%	AIHA J. 47(2):135–137 (1986)	(4)
Amyl acetate	Charcoal tube	Gas chromatography, FID	5.1%	NIOSH: 1450 (3rd Ed.)	(5)
Aniline	Silica gel tube	Gas chromatography, FID	6.0%	NIOSH: 2002 (3rd Ed.)	(6)
Antimony compounds	CE filter	Atomic absorption, flame	5.9%	NIOSH: Sa (Vol. 2, 2nd Ed.)	(7)
Arsine	Charcoal tube	Atomic absorption, graphite furnace	8.7%	NIOSH: 6001 (3rd Ed.)	(8)
Asbestos	CE filter	Microscope, counting	25%	NIOSH: 7400 (3rd Ed.), Federal Register, Vol. 51, No. 119, EPA 560/5-85-024	(9)
Barium compounds	CE filter	Atomic absorption, flame	5.8%	NIOSH: 173 (Vol. 5, 2nd Ed.)	(10)
Benzene	Charcoal filter	Gas chromatography, FID	5.9% CT	NIOSH: 1501 (3rd Ed.)	(11)

acetic acid using solid sorbents as an alternative to impinger methods because impingers restrict workers mobility and are subject to spillage [9]. The use of these tubes that contain firebrick (Chromosorb P) impregnated with 5% sodium carbonate has greatly facilitated sampling for these acids; these tubes are compatible with ion chromatography analysis methodology, thereby providing improved sensitivity, selectivity, and reliability. Chromosorb P tubes for acids are commercially available.

The prevalent use of ethylene oxide (ETO) and the lowering of the health standard to 1 ppm caused USAEHA to explore a new method for its sampling and analysis. A procedure

---

INDUSTRIAL HYGIENE AIR SAMPLE DATA

For use of this form see USAEHA TG 141; the proponent is HSHB-LO.

Return Address *(complete address including Zip Code)*   **Point of Contact** *(name/AUTOVON)*

Associated Bulk Samples
☐ Yes   ☐ No

Samples Collected By | **Date Collected** | **Date Shipped**

Bulk Sample No(s):

Project Number | **Sampled Installation** | **ARLOC**

Location *(BLDG/AREA)* | **Description of Operation** *(details on reverse)*

☐ **Persons Exposed**   ☐ **Hrs/Day** | **Method of Collection**

Associated Complaints *(be specific)* *(state NONE if applicable)*

Analysis Desired

Sampling Data

Sample No.							
Pump No.							
Time On							
Time Off							
Total Time (min)							
Flow Rate (LPM)							
Volume (Liters)							
GA/BZ							
Employee Name/ID							
Laboratory No.							

B L A N K

Results

Comments to Lab:

Lab Use Only

Analyst *(initials)*	Reviewed By *(initials)*	Date Received	Date Dispatched

AEHA Form 9-R, 1 Oct 84

Replaces AEHA Form 9, 1 Oct 80 which is obsolete.

FIG. 1—*Industrial hygiene air sample form (front).*

was developed based upon the entrapment and reaction of ETO on a chemically impregnated air sampling tube to form 2-bromoethanol, with subsequent analysis by gas chromatography (GC) with electron capture detector (ECD) [10]. These tubes are also commercially available.

The U. S. Army has the requirement to perform industrial hygiene sampling for trinitrotoluene (TNT) and cyclonite (RDX). The USAEHA chemists have developed [11] a procedure for these explosives that involves collection of the vapors on commercial tenax tubes, solvent desorption, and analysis by GC with an electron capture detector.

The low Occupational Safety and Health Administration (OSHA) health standard for hexamethylene diisocyanate (HDI) and its large usage in Army painting operations created

Calibration Information				
Pump No.	Calibration (L/min) Pre-Use	Post-Use	Rotometer Setting	Date
			Name of Calibrator	

Operation

Source of Contaminant:

Operation Employee(s) Perform:

Ventilation: ☐ Local Exhaust   ☐ General Area   ☐ None

Personal Protective Equipment *(check if worn)*

☐ Respiratory Protective Equipment   Type: _____

☐ Protective Clothing   Type: _____

☐ Gloves   Type: _____

☐ Goggles/Face Shield

☐ Ear Protection

☐ Other: _____

Field Notes/Additional Comments

FIG. 2—*Industrial hygiene air sample form (back).*

the need to develop a new analytical procedure for HDI [12]. After collection of the isocyanate in the acidic absorbing solution recommended by Marcali, the sample is fluoroacylated and analyzed by GC with electron capture detector. A similar procedure was developed for analysis of diphenylmethane diisocyanate (MDI) and toluene diisocyanate (TDI) [13]. The procedures for all three isocyanates have the advantage of specificity, low limits of detection, and excellent accuracy and precision.

BULK SAMPLE DATA

For use of this form see USAEHA TG 141; the proponent is HSHB-LO.

Return Address *(complete address including Zip Code)*	Point of Contact *(name/AUTOVON)*

Sampled Installation	Project Number	ARLOC

Samples Collected By	Date Collected	Date Shipped

Description of Operation	Location *(BLDG/AREA)*

Associated Complaints *(be specific)*

Associated Air Samples ☐ Yes ☐ No   If yes, list sample numbers

Label Information

Trade Name	NSN	Manufacturer
Address		MSDS Attached ☐ Yes ☐ No

Analysis Desired

Lab Use Only	Sample No.	Constituents	Results	Remarks

Comments to Lab:

Lab Use Only

Analyst *(initials)*	Reviewed By *(initials)*	Date Received	Date Reported
Procedures Performed	Comments:		

AEHA Form 8, 1 Oct 84

Replaces AEHA Form 8, 1 Oct 80 which is obsolete.

FIG. 3—*Industrial hygiene bulk sample form.*

Sample Collection Forms

The IH Sampling Guide provides administrative details such as the forms required for sample submission (Figs. 1 through 3) and instructions for filling out the forms. The procedures for shipping air or bulk samples to USAEHA or other supporting U. S. Army laboratory are also presented. These administrative details facilitate processing and analysis of industrial hygiene samples.

Conclusion

The state of the art in sampling and analysis and the TLVs change periodically, consequently, the IH Sampling Guide is updated every two years [14]. Significant changes are provided to the field personnel through the "Preventive Medicine Information Letter," a quarterly publication of the U. S. Army Health Services Command, or by distribution of pertinent changes to all appropriate activities or both. The Industrial Hygiene Sampling Instructions, Technical Guide-141, is approved for public release with distribution unlimited. The IH Sampling Guide is to be incorporated into the U. S. Army Industrial Hygiene Evaluation Guide that is being written at USAEHA.

References

[1] Mazur, J. F., Podolak, G. E., Esposito, G. G., Rinehart, D. S., and Glenn, R. E., *Journal, American Industrial Hygiene Association*, Vol. 41, 1980, pp. 317–321.
[2] Mazur, J. F., Rinehart, D. S., Esposito, G. G., and Podolak, G. E., *Journal*, American Industrial Hygiene Association, Vol. 42, 1981, pp. 752–756.
[3] *Criteria for a Recommended Standard—Occupational Exposure to Waste Anesthetic Gases and Vapors*, HEW Publication No. 77–140, Dept. of Health, Education, and Welfare, National Institute for Occupational Safety and Health, U. S. Government Printing Office, Washington, DC, 1977.
[4] *Documentation of the Threshold Limit Values*, 5th ed, American Conference of Government Industrial Hygienists, Cincinnati, 1986, p. 523–525.
[5] Charell, P. R. and Hawley, R. E., *Journal*, American Industrial Hygiene Association, Vol. 42, 1980, pp. 353–360.
[6] *Criteria for a Recommended Standard—Occupational Exposure to Sodium Hydroxide*, HEW Publication No., 76-105, Dept. of Health, Education, and Welfare, National Institute for Occupational Safety and Health, U. S. Government Printing Office, Washington, DC, 1975.
[7] *Manual of Analytical Methods*, 2nd ed., Vols. 1–7, National Institute for Occupational Safety and Health, Publication Nos. 77–147A, 77–157B, 77–157C, 78–175, 79–141, 80–125, 82–100, Cincinnati, 1977–1982.
[8] *Manual of Analytical Methods*, 3rd ed., P. Eller, Ed., DHHS (NIOSH) Publication No. 84–100, Department of Health and Human Services, Cincinnati, 1984.
[9] Williams, K. E., Esposito, G. G., and Rinehart, D. S., *Journal*, American Industrial Hygiene Association, Vol. 42, 1981, pp. 476–478.
[10] Esposito, G. G., Williams, K., and Bongiovanni, R., *Analytical Chemistry*, Vol. 56, No. 11, 1984, pp. 1950–1953.
[11] Bishop, R. W., Ayers, T. A., and Rinehart, D. S., *Journal*, American Industrial Hygiene Association, Vol. 42, 1981, pp. 586–589.
[12] Esposito, G. G. and Dolzine, T. W., *Analytical Chemistry*, Vol. 54, 1982, pp. 1572–1575.
[13] Bishop, R. W., Ayers, T. A., and Esposito, G. G., *Journal*, American Industrial Hygiene Association, Vol. 44, No. 3, 1983, pp. 151–155.
[14] *Threshold Limit Values and Biological Exposure Indices for 1986–1987*, American Conference of Governmental Industrial Hygienists, Cincinnati, 1986.

Elmer S. McKee[1] and Paul W. McConnaughey[1]

Detector Tubes

REFERENCE: McKee, E. S. and McConnaughey, P. W., **"Detector Tubes,"** *Sampling and Calibration for Atmospheric Measurements, ASTM STP 957,* J. K. Taylor, Ed., American Society for Testing and Materials, Philadelphia, 1987, pp. 176–189.

ABSTRACT: The specificity of various detector tube systems is discussed. Some chemical reactions are given for various types of tubes to illustrate specific and nonspecific tubes. Most detector tube systems are not specific and this is one of the major disadvantages of using detector tubes. The accuracy of detector tubes is also discussed. Two methods of estimating accuracy are illustrated: (1) using MIL-STD 414, which is the method that the National Institute for Occupational Safety and Health (NIOSH) used for certifying detector tubes; and (2) the "DuPont" method, where the bias and precision (mean coefficient of variance) are estimated, and the overall accuracy is obtained from these values. Approximate accuracies of some detector systems are given. Finally, some advantages and disadvantages of detector tubes are listed.

KEY WORDS: air quality, calibration, sampling, atmospheric measurements, accuracy, specificity, chemical reactions, detector tubes, toxic gas monitoring, analysis

Detector tubes have been used to detect toxic substances for many years. The beginning of the use of colorimetric detector tubes was at Harvard University in 1917, where a tube was developed to detect carbon monoxide. One of the first commercial detector tubes offered was A Mine Safety Appliances Co. (MSA) carbon monoxide detector tube that was included in a 1929 MSA catalog. It is one of the first references to a detector tube for commercial use. Today, there are four main manufacturers of detector tubes.[2] In 1935, Littlefield, Yant, and Berger published a report for the U. S. Bureau of Mines on a hydrogen sulfide tube.[3] Since then, colorimetric detector tubes have been developed for many substances.

A quote from an American Industrial Hygiene Association's publication[4] is a good introduction to this subject.

Perhaps the most widely used detection techniques by industrial hygienists, safety engineers, safety specialists, and others has been the colorimetric detection devices, commonly referred to as gas indicator or detector tubes. Their simplicity, low initial cost and versatility regarding detection of numerous contaminants make them a popular instrument for field use. Nevertheless, like all instruments, these devices have limitations regarding applicability, specificity, and accuracy. The user must be familiar with these limitations if he is to make proper judgements.

---

[1] Manager, Chemical Research and Development, and retired, respectively, Mine Safety Appliances Company, Pittsburgh, PA 15230.

[2] Manufacturers of detector tubes include: Dragerwerk AG, West Germany; Gastec, Japan; Kitagawa, Japan; and Mine Safety Appliances Co., Pittsburgh, PA.

[3] Littlefield, J. B., Yant, W. P., and Berger, L. B., "A Detector for Quantitative Estimation of Low Concentration of Hydrogen Sulfide," U. S. Bureau of Mines, Pittsburgh, 1935.

[4] *Direct Reading Colorimetric Indicator Tubes Manual,* First Edition, American Industrial Hygiene Association, Akron, OH, 1976.

Although this statement was made nine years ago, it is still applicable today.

Detector tubes are usually used to measure toxic substances in industrial atmospheres in the parts per million (ppm) range and not for U. S. Environmental Protection Agency (EPA) environmental studies in the parts per billion (ppb) range.

This paper will deal mainly with the specificity and accuracy of detector tubes.

Specificity

General

Detector tubes are based on chemical reagent systems so that the reagent system in a detector tube constitutes a ready-made analysis. Today, detector tubes are used to detect many substances. A new American Society for Testing and Materials (ASTM) practice, ASTM Standard Practice for Measuring the Concentration of Toxic Gases or Vapors Using Detector Tubes D 4490-85, is in the process of being published. It lists over 200 substances that can be detected by detector tubes; however, one of the big drawbacks of detector tubes is their lack of specificity. For example, MSA lists about 150 different substances that can be detected with its tubes; and yet, MSA uses only about 50 different chemical indicators, pointing out the fact that the same chemical can be used to detect a number of different chemical compounds, indicating a lack of specificity of some systems. This is probably one of the biggest disadvantages of detector tubes.

To better understand this lack of specificity, a look at some of the chemical reactions that are used in detector tubes would be appropriate.

Chemical Reactions Used in Detector Tubes

Use is made of all the classical types of chemical reactions in detector tubes: oxidation-reduction, acid-base, substitution, addition, etc. A few specific reactions follow.

Oxidation-Reduction Reactions—This was the type of reaction that was used by Hoover and Lamb at Harvard University in 1917 to detect carbon monoxide. The procedure was patented in 1919. The material was called Hoolamite, after its inventors: Hoo(ver) Lam(b)ite. The indicator used a mixture of iodine pentoxide and fuming sulfuric acid on pumice. In the reaction, the iodine is reduced and the carbon monoxide is oxidized, thus

$$I_2O_5 + 5CO \xrightarrow{\text{H}_2\text{S}_2\text{O}_7} 5CO_2 + I_2$$

The color change is from white to brown/green.

Some other oxidants used in detector tubes are also shown in Table 1.

As can be seen, and easily understood from elementary chemistry, the iodine pentoxide and dichromate oxidation-reduction reactions are nonspecific, although the sensitivities of indication are different for different compounds. For instance, I_2O_5 + fuming sulfuric acid gives a stain of 14.5 mm for 50 ppm benzene using one pump stroke, 10.5 mm for toluene, and 7.5 mm for xylene. But if these three compounds were together in a mixture, using detector tubes, there is no way of telling what the concentration of each is in the mixture.

The reaction of carbon monoxide (CO) with potassium palladosulfite is fairly specific. For instance, 10 ppm H_2S in 50 ppm CO causes about a 10% increase in the length of stain as does 1 ppm acetylene, but in most circumstances these substances would not be present in measurable concentrations when testing for CO. The molybdate reaction will oxidize and split double bonds, but other strongly reducing materials can interfere.

TABLE 1—*Oxidation/reduction reactions used in detector tubes.*

Reaction		Color Change	Used For
$I_2O_5 + H_2S_2O_7$ $\xrightarrow{SeO_2}$	$I_2 + Ox.Pr^a$	white/brown-green	CO, many organics
$I_2O_5 + H_2S_2O_7 \rightarrow$	$I_2 + Ox.Pr$	white/brown-green	CO, many organics
$I_2O_5 + H_2SO_4 \rightarrow$	$I_2 + Ox.Pr$	white/brown-green	aromatic hydrocarbons
$Cr_2O_7 + H_2SO_4 \rightarrow$	$Cr^{+3} + Ox.Pr$	orange/green	aliphatic hydrocarbons
$Cr_2O_7 + H_3PO_4 \rightarrow$	$Cr^{+3} + Ox.Pr$	orange/green	alcohols

Special:
$CO + K_2Pd(SO_3)_2 \rightarrow CO_2 + SO_2 + K_2SO_3 + Pd$, yellow/brown, CO-specific
$R_1C = CR_2 + Mo^{+6} \rightarrow Mo^{+3} + Ox.Pr$ (splitting of double bond), yellow/blue, unsaturated hydrocarbons

^{a} Ox.Pr = oxidation product.

Acid-Base Reactions—Many detector tubes use acid-base indicators impregnated on a neutral or buffered substrate, to monitor gaseous substances that have acidic or basic properties. Some of the compounds for which acid-base indicators are used are: acetic acid, ammonia and amines, carbon dioxide, hydrazine, hydrogen chloride, hydrogen fluoride, nitric acid, and sulfur dioxide. A few examples are given in Table 2.

As with tubes using oxidation-reduction reactions, acid-base detector tubes are not specific. Other acids and bases will interfere with the detection, particularly strong acids and bases.

Other Types of Reactions—Some other types of reactions used in detector tubes are shown in Table 3. The classical substitution reaction of hydrogen sulfide with a lead salt is one of the more stable, specific, and accurate reactions used. The color change is from white to grey/black, depending on the challenge concentration of hydrogen sulfide and the consequent amount of lead sulfide formed. Another substitution reaction is the reaction of carbonyl compounds with dinitrophenyl hydrazine, with the splitting off of a molecule of water. This reaction is fairly specific for the carbonyl class of compounds.

The reaction of chlorine with tetraphenylbenzidine, $(C_6H_5)_2NC_6H_4C_6H_4N(C_6H_5)_2$, is a good example of an addition reaction. The chlorine is taken up by the tetraphenylbenzidine to form a quinoidimonium salt. This reaction is fairly specific, although other halogens and nitrogen dioxide cause some interference.

A two-stage reaction occurs between benzene, formaldehyde, and sulfuric acid. First, the benzene and formaldehyde combine to form diphenylmethane and a molecule of water.

TABLE 2—*Acid/base indicators used in detector tubes.*

Challenge Gas/Vapor	Indicator	Color Change
Acetic acid	phenol red	pink to white
Ammonia	bromophenol blue or thymol blue	orange to blue/green; lavender to pale yellow
Carbon dioxide	thymol blue	blue to pale yellow
Hydrazine	bromophenol blue	yellow to blue
Hydrogen chloride	bromophenol blue or congo red	blue to white; pink to blue
Sulfur dioxide	phenol red or bromocresol green	pink to white; blue to yellow

TABLE 3—*Other types of reactions used in detector tubes.*

Reaction	Color Change	Specificity
Substitution		
$H_2S + PbAc_2 \rightarrow PbS + 2HAC$	white/black	specific
$R_1R_2C = O + H_2NNHC_6H_3(NO_2)_2 \rightarrow R_1R_2C = NNHC_6H_3(NO_2)_2 + H_2O$	pale yellow/yellow	specific for carbonyls
dinitrophenylhydrazine $\rightarrow$ hydrazone		
Addition		
$Cl_2 + TPB = Cl{-}TPB{-}Cl$	white/blue	fairly specific
tetraphenylbenzidine $\rightarrow$ quinoidimonium salt		
Combination		
$2C_6H_6 + CH_2O \rightarrow C_6H_5CH_2C_6H_5 + H_2O$	white/brown	specific for aromatics (or aldehydes)
$+$ $2H_2SO_4$ $\xrightarrow{}$ $C_6H_5CH = \bigcirc = O + 3H_2O = 2SO_2$		
paraquinoid compound		
Use of strong oxidant to split off detectable substance		
Chlorinated hydrocarbons $+ KMnO_4 + H_2SO_4$ (conc.) $\rightarrow Cl_2$		
$Cl_2 + TPB \rightarrow Cl{-}TPB{-}Cl$	white/blue	nonspecific
Use of heat to break down compound into detectable substance		
$CCl_4 + $ heat $\rightarrow Cl_2$ use Cl_2 tube	white/blue	nonspecific
$CH_3CN + $ heat $\rightarrow NO_2$ use NO_2 tube	white/blue	nonspecific
$CF_2Cl_2 + $ heat $\rightarrow Cl_2$ use Cl_2 tube	white/blue	nonspecific

Then, the sulfuric acid oxidizes the diphenylmethane to a paraquinoid compound, which has a brownish color. This reaction can be used for the detection of either benzene or formaldehyde. The reaction is nonspecific, as other aromatics and aldehydes will also react in this manner.

For a large number of gases and vapors, because of their stability, it is not possible to obtain directly a color indication with detector tube reagent systems. In many instances, a solution is found by chemically or thermally breaking down the compound into a detectable substance. Some examples are shown in Table 3. Chlorine can be split off of many chlorinated hydrocarbons by using the powerful oxidant, potassium permanganate, and concentrated sulfuric acid. The permanganate and acid are kept in separate ampoules until they are to be used. At this time, the ampoules are broken and the contents mixed together. The gas containing the chlorinated hydrocarbon is drawn through this oxidizing bed, releasing chlorine, which is detected with the chlorine detector tube.

With more stable compounds, more drastic treatment is needed to break them down into a detectable substance. Here a pyrolizer is used, which heats the compounds in air to 400 to 500°C. Under these conditions, detectable fragments are given off by the compounds. Some examples are: carbon tetrachloride releasing chlorine, freon 12 (dichlorodifluoromethane) also giving off chlorine, and acetonitrile yielding nitrogen dioxide. The chlorine and nitrogen dioxide are readily detected by the respective detector tubes.

Here again, heat will break down many organic nitrogen compounds, yielding nitrogen dioxide, or chlorinated compounds, yielding chlorine, so the detection is not specific. Also, strong oxidants will break down many halogenated hydrocarbons to chlorine and other products. The sensitivity of detection of the various compounds varies considerably, but the overall detection is nonspecific.

Thus, it can be seen that detector tubes must be used with judgment and the results carefully interpreted to prevent invalid conclusions.

Accuracy

There are many things that influence the accuracy of detector tubes, but before we discuss these, let us consider two methods for determining accuracy: (1) using MIL-STD 414, and (2) using a laboratory validation system developed by DuPont. Actually, the MIL-STD 414 method is more of an acceptance test, but some inference of accuracy can be made from it.

Method using MIL-STD 414

This is the method that the National Institute for Occupational Safety and Health (NIOSH) used in their detector tube certification program, and it was published in the *Federal Register* in 1973.[5] The method to determine lot acceptance consists of the following steps:

1. Use MIL-STD 414, Level II, acceptable quality level (AQL) 6.5%, double specific limit, variables unknown.
2. Test at 1/2, 1, 2, and 5 times the threshold limit value (TLV).
3. Tolerance allowed: ±25% at 1, 2, and 5 TLVs; ±35% at 1/2 TLV.
4. Sample size determined from MIL-STD 414, Tables A-2 and B-3 (see Table 4).
5. Determine $\overline{X}$, σ, U (upper specific limit), and L (lower specific limit) for each concentration.

[5] *Federal Register,* Vol. 18, No. 89, Title 42, Part 84, U. S. Government Printing Office, Washington, DC, 8 May 1973.

TABLE 4—*Excerpt from Tables A-2 and B-3, MIL-STD 414.*

Lot Size	Sample Size for Level II
501 to 800	10
801 to 1 300	15
1 301 to 3 200	20
3 201 to 8 000	25
8 001 to 22 000	30
22 001 to 110 000	35

6. Determine Q_U (quality index, upper limit) $= (U - \overline{X})/\sigma$ and determine Q_L (quality index, lower limit) $= (\overline{X} - L)/\sigma$.

7. Calculate P_U (percent of tubes with readings greater than U) and P_L (percent of tubes with readings less than L) from MIL-STD 414, Error Table B-5 (see Table 5).

8. If $P_U + P_L$ is less than the maximum allowable error, M, (MIL-STD 414, Error Table B-3), the lot is acceptable (see Table 6).

where

$\overline{X}$ = the average measured concentration as measured by the tubes;

σ = standard deviation of the measured concentration;

U = upper specification limit. At 1, 2, and 5 times the TLV, $U = X + 0.25X$. At 1/2 TLV, $U = X + 0.35X$, where X = the true test concentration; and

L = lower specification limit. At 1, 2, and 5 times the TLV, $L = X - 0.25X$. At 1/2 TLV, $L = X - 0.35X$.

Several examples will help to clarify the use of this method. Consider Lot 30 of hydrogen sulfide tubes, Part No. 460058. There were 36 846 tubes in this lot. From MIL-STD 414, Level II, Tables A-2 and B-3, the number of tubes to be tested at each concentration is 35 (see Table 4). The test results are shown in Table 7. Each of the millimetre readings is an average of three readers. The millimetre readings are converted to ppm, using Fig. 1. Actually, today the ppm is read directly from the scale on the tubes. Then, applying MIL-STD 414, Level II, AQL 6.5%, these results were analyzed and gave the values shown in Table 8 [where U at 1/2 TLV $= 5.0 + 0.35(5.0) = 6.75$; $L = 5.0 - 0.35(5.0) = 3.25$; Q_U

TABLE 5—*Excerpt from Table B-5, MIL-STD 414.*[a]

Q_U or Q_L	Sample Size				
	15	20	25	30	35
1.10	13.51	13.52	13.52	13.53	13.54
1.25	10.34	10.40	10.43	10.46	10.47
1.50	6.20	6.34	6.41	6.46	6.50
1.75	3.37	3.56	3.66	3.72	3.77
2.00	1.62	1.81	1.91	1.98	2.03
2.25	0.660	0.816	0.905	0.962	1.002
2.50	0.214	0.317	0.380	0.421	0.451
3.00	0.006	0.025	0.042	0.055	0.065

$$Q_U = \frac{U - \overline{X}}{\sigma} \qquad Q_L = \frac{\overline{X} - L}{\sigma}$$

[a] Estimation of the percentage of detector tubes (P_U or P_L) with an error above or below the specification limit (U or L).

TABLE 6—*Excerpt from Table B-3, MIL-STD 414.*

Sample Size	Acceptable Quality Levels (normal inspection), M					
	1.00	1.50	2.50	4.00	6.50	10.00
3	...	...	7.59	18.86	26.94	33.69
4	1.53	5.50	10.92	16.45	22.86	29.45
5	3.32	5.83	9.80	14.39	20.19	26.56
7	3.55	5.35	8.40	12.20	17.35	23.29
10	3.26	4.77	7.29	10.54	15.17	20.74
15	3.05	4.31	6.56	9.46	13.71	18.94
20	2.95	4.09	6.17	8.92	12.99	18.03
25	2.86	3.97	5.97	8.63	12.57	17.51
30	2.83	3.91	5.86	8.47	12.36	17.24
35	2.68	3.70	5.57	8.10	11.87	16.65
40	2.71	3.72	5.58	8.09	11.85	16.61
50	2.49	3.45	5.20	7.61	11.23	15.87
75	2.29	3.20	4.87	7.15	10.63	15.13
100	2.20	3.07	4.69	6.91	10.32	14.75
150	2.05	2.89	4.43	6.57	9.88	14.20
200	2.04	2.87	4.40	6.53	9.81	14.12

at 1/2 TLV = $(6.75 - 5.42)/0.386 = 3.44$; and P_U, corresponding to this Q_U from Table B-5 in MIL-STD 414 = 0.007 (see Table 5)]. Similarly, Q_L at 1/2 TLV = $5.42 - 3.25/0.386 = 5.62$, $P_L = 0.0051$, and $P_U + P_L = 0.012$. The value of M, the maximum allowable error, from Table B-3 MIL-STD 414, corresponding to a sample size of 35 and an AQL of 6.50 is 11.87 (see Table 6). Since $P_U + P_L$ is less than this, the lot is acceptable at this concentration. From the $P_U + P_L$ values at other concentrations, Table 8, the lot is acceptable at all test concentrations.

Another example of using MIL-STD 414 to estimate accuracy of detector tubes is Lot 18, toluene tubes, Part No. 461371. This lot had 3952 tubes, requiring a sample size of 25. The data for this lot is given in Table 9. The Calibration Graph, to convert millimetres to ppm, is shown in Fig. 2. The analysis for this lot is shown in Table 10. Again, all values of $P_U + P_L$ are less than M, although note that at the TLV (100 ppm) the values are very close, indicating that the accuracy here is about $\pm25\%$, given a 6.5 AQL.

This leads to another useful estimation of accuracy using MIL-STD 414. The acceptance of a lot is based on $P_U + P_L$ being less than M. The values of U and L, the specification limits, can be adjusted so that $P_U + P_L = M$. This then would be the lowest percentage variance for which the tubes would meet the criteria of MIL-STD 414 at the specified AQL. As an example, consider the data for toluene, Lot 18, in Table 10. At 1/2 TLV, $\overline{X} = 54.47$ and $\sigma = 7.52$, and Q_U and Q_L are adjusted so that the sum of their corresponding P values just equals M. Since Q_U and Q_L depend on the upper and lower specific limits, U and L, and these values depend on the allowable percentage error, the allowable percentage error value that satisfies the condition $P_U + P_L = M$ is the smallest tolerance that the lot can have under the test conditions. For example, at 1/2 TLV, suppose the specific limits are set at $\pm25\%$ instead of 35%. Then $U = 50 + 0.25(50) = 62.5$ instead of 67.5 and $Q_U = (62.5 - 54.47)/7.52 = 1.07$ and $P_U = 14.21$. Where P_U, itself, is larger than M, the lot would not pass at $\pm25\%$ at 1/2 TLV. By trial and error, the lot at 1/2 TLV was found to pass at $\pm26.5\%$ as shown in Table 10, Section 4. At 1 TLV, the tolerance limits were found to be just under 25%, $\pm24.7\%$. At 2 and 5 TLVs, the tolerances are $\pm18.0\%$ and $\pm12.1\%$, respectively.

Using this same technique for H_2S Lot 30, the acceptable tolerances at a 6.5% AQL at 1/2, 1, 2, and 5 TLVs are estimated to be 17.7%, 18.59%, 13.6%, and 11.5%, respectively (see Table 8, Section 4). With these two lots, as the concentration increases, the accuracy improves. In general, the accuracy of detector tubes improves at higher concentrations.

Method using the DuPont System

This method to determine accuracy consists of the following steps:

1. Test at least six tubes at 1/2, 1, 2, and 5 times the TLV.
2. For each concentration, calculate $\overline{X}$ and σ.

TABLE 7—*Data for Lot 30 H_2S tubes, Part No. 460058.*

Tube Number	5 ppm, 6 PSs		10 ppm, 6 PSs		20 ppm, 6 PSs		50 ppm, 2 PSs	
	mm	ppm	mm	ppm	mm	ppm	mm	ppm
1	14.0[a]	4.93	19.0[a]	7.62	38.0[a]	20.48	31.7[a]	53.93
2	13.3	4.60	20.8	8.69	36.5	19.34	30.5	51.15
3	13.3	4.60	21.7	9.19	39.0	21.26	28.7	46.86
4	15.5	5.70	21.8	9.29	39.2	21.39	30.7	51.54
5	15.3	5.61	19.5	7.91	36.8	19.59	30.5	51.15
6	15.3	5.61	19.0	7.62	39.2	19.84	30.8	51.94
7	14.8	5.35	22.8	9.90	39.5	21.65	31.0	52.34
8	16.0	5.96	19.8	8.10	38.3	20.74	33.3	57.98
9	15.0	5.44	17.7	6.87	29.2	14.04	29.7	49.19
10	16.0	5.96	22.7	9.80	36.0	18.96	28.3	46.10
11	14.7	5.18	23.0	10.01	39.3	21.52	28.7	46.86
12	15.3	5.61	22.2	9.49	39.2	21.39	30.2	50.36
13	14.8	5.35	23.0	10.01	39.0	21.26	30.3	50.76
14	14.2	5.01	21.5	9.09	38.3	20.74	29.8	49.58
15	14.7	5.27	19.3	7.81	39.8	21.91	29.8	49.58
16	15.8	5.87	21.0	8.79	38.0	20.48	30.2	50.36
17	13.3	4.60	20.7	8.59	40.8	22.70	30.3	50.76
18	13.7	4.76	21.0	8.79	36.0	18.96	30.0	49.97
19	16.3	6.14	22.0	9.39	34.3	17.72	29.8	49.58
20	15.2	5.52	21.2	8.89	40.3	22.30	30.3	50.76
21	15.0	5.44	21.0	8.79	40.8	22.70	32.0	34.34
22	15.7	5.78	22.5	9.70	39.8	21.91	30.5	51.15
23	15.0	5.44	22.3	9.59	39.5	21.39	31.3	53.13
24	15.2	5.52	21.8	9.29	38.9	21.00	31.2	52.73
25	15.2	5.52	21.7	9.19	40.3	22.30	31.8	54.33
26	15.5	5.70	22.0	9.39	40.3	22.30	30.2	50.36
27	14.2	5.01	21.0	8.79	39.5	21.65	27.8	44.95
28	15.0	5.44	22.3	9.59	38.0	20.48	28.6	46.48
29	15.7	5.78	22.7	9.80	38.3	20.74	29.0	47.64
30	15.3	5.61	22.0	9.39	35.7	18.71	30.2	50.36
31	14.7	5.27	21.5	9.09	36.5	19.34	30.7	51.54
32	15.7	5.78	22.5	9.70	38.7	21.00	29.8	49.58
33	15.0	5.44	22.7	9.80	38.0	20.48	30.3	50.76
34	15.2	5.52	22.8	9.90	36.3	19.21	29.2	48.02
35	15.0	5.44	22.3	9.59	37.3	19.72	29.2	48.02
$\overline{X}$		5.42		9.07		20.55		49.83
σ		0.386		0.768		1.644		3.655

[a] These values are the average of three tube readers.

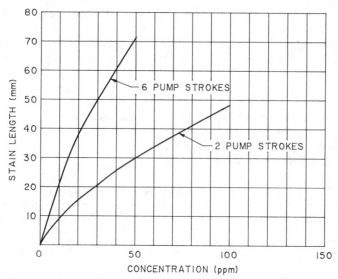

FIG. 1—*Calibration curves for H₂S detector tubes, Lot 30.*

3. From these, calculate the coefficient of variance (*CV*) and the bias (*b*) at each concentration.

$$CV = \sigma/\overline{X}, \qquad b = \frac{\overline{X} - X}{X}$$

where X = true concentration.

TABLE 8—*Analysis of H₂S detector tubes, Lot 30, using MIL-STD 414.*

	1/2 TLV (5 ppm)	1 TLV (10 ppm)	2 TLV (20 ppm)	5 TLV (50 ppm)
1. Number of tubes in lot = 36 846. Using MIL-STD 414, Table A-2, Level II, sample size is 35.				
2.				
U	6.75	12.5	25.0	62.5
L	3.25	7.5	15.0	37.5
$\overline{X}$	5.42	9.07	20.55	49.83
σ	0.386	0.768	1.644	3.655
Q_U	3.44	4.47	2.71	3.47
Q_L	5.62	2.04	3.37	3.37
P_U (Table B-5)	0.007	0.005	0.220	0.007
P_L (Table B-5)	0.005	1.82	0.011	0.011
$P_U + P_L$	0.012	1.825	0.231	0.018
M (Table B-3)	11.87	11.87	11.87	11.87
3. $P_U + P_L$ is less than the maximum allowable error at all concentrations; therefore, the lot is accepted.				
4. Least ± percentage tolerance for which the lot would be accepted.				
%	17.7	18.5	13.6	11.5
Q_U	1.20	3.62	1.32	1.62
Q_L	3.38	1.20	1.99	1.53
P_U	11.43	0.005	9.22	5.16
P_L	0.01	11.63	2.13	6.23
$P_U + P_L$	11.44	11.635	11.35	11.39
M	11.87	11.87	11.87	11.87

TABLE 9—*Data for Lot 18, toluene tubes, Part No. 461371.*

Tube Number	50 ppm, 3 PSs		100 ppm, 3 PSs		200 ppm, 3 PSs		500 ppm, 2 PSs	
	mm	ppm	mm	ppm	mm	ppm	mm	ppm
1	15.50	47.8	23.50	106.0	33.33	206.6	43.33	490.3
2	15.50	47.8	24.33	113.2	28.83	156.6	38.00	383.4
3	16.17	51.9	19.50	74.2	36.00	239.4	43.67	497.5
4	12.83	33.3	17.67	61.5	28.50	153.2	42.83	478.7
5	18.00	63.7	20.83	84.2	29.17	160.1	41.17	445.5
6	17.50	6.03	22.83	100.3	33.83	212.6	45.33	533.6
7	16.83	56.0	22.00	93.4	33.00	202.7	42.83	479.7
8	16.33	52.9	23.50	106.0	32.67	198.9	41.83	459.0
9	17.17	58.2	17.83	62.5	30.83	178.0	42.67	476.4
10	17.00	57.1	21.83	92.0	33.17	204.7	45.67	541.1
11	15.83	49.8	24.83	117.7	33.33	206.6	44.67	519.1
12	17.50	60.3	20.00	77.9	35.50	233.1	45.17	530.0
13	16.83	56.0	24.17	111.8	30.67	176.24	42.00	462.5
14	18.00	63.7	22.00	93.4	35.50	233.1	46.00	548.4
15	18.50	67.1	21.67	90.8	34.67	222.8	46.33	555.8
16	16.33	52.9	20.67	82.9	33.83	212.6	44.33	511.7
17	16.33	52.9	23.17	103.1	33.17	204.7	44.83	522.6
18	16.67	55.0	23.50	105.0	33.83	212.6	42.67	476.4
19	16.00	50.8	24.17	111.8	35.00	226.8	42.33	469.3
20	17.00	57.1	23.50	106.0	33.00	202.7	44.00	504.6
21	18.17	64.8	23.83	108.8	32.00	214.6	44.17	508.3
22	16.67	51.9	24.67	116.3	33.33	206.6	43.67	497.5
23	16.17	55.0	22.17	94.8	30.00	169.0	44.67	519.1
24	14.00	39.4	20.50	81.6	31.50	185.5	47.00	571.0
25	16.83	56.0	23.17	103.1	32.83	200.7	43.33	490.3
$\overline{X}$		54.47		45.97		200.8		498.9
σ		7.52		15.89		28.91		40.00

4. Calculate the mean coefficient of variance (*MCV*) and the absolute mean bias ($/\overline{b}/$).

$$MCV = \sqrt{\frac{\Sigma (n_i - 1) CV_i^2}{\Sigma (n_i - 1)}}$$

$$/\overline{b}/ = \frac{\Sigma n_i b_i}{\Sigma n_i}$$

where n_i = number of observations at the *i*th concentration.
5. Calculate the overall accuracy (OA), that is

$$\pm [/\overline{b}/ + 2(MCV)]100$$
$$\text{bias} + \text{precision}$$

Using the same data from Table 7 for H_2S Lot 30, the statistical analysis is shown in Table 11. Using this method of determining accuracy, the overall accuracy from 1/2 to 5 times the TLV is ±15.9%. This is in line with the accuracy estimated for this lot using MIL-STD 414; however, if the accuracy values estimated at each concentration are used, the DuPont method gives poorer values.

Going through the same analysis and comparison for the toluene tubes of Lot 18 indicates

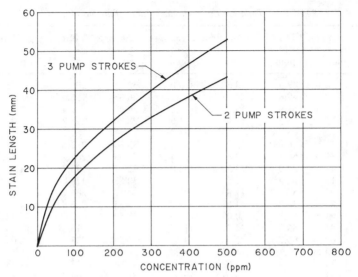

FIG. 2—*Calibration curves for toluene detector tubes, Lot 18.*

a significant difference for the two systems (see Table 12). The MIL-STD 414 method accepted the lot, though barely at 1 TLV, while the DuPont method would reject the lot. The estimated accuracy by the latter method is ±27.1%. Of course, this analysis is over the whole range of TLVs from 1/2 to 5, whereas the MIL-STD method considers the variance at each individual concentration. However, the estimated errors at the individual concentrations are all higher by the DuPont system: 36.5%, 37.2%, 24.4%, and 16.2% versus 26.5%, 24.7%, 18.0%, and 12.1% for 1/2, 1, 2, and 5 TLVs, respectively. The MIL-STD

TABLE 10—*Analysis of toluene detector tubes, Lot 18, using MIL-STD 414.*

1. Number of tubes in lot = 3952. Using MIL-STD 414, Table A-2, Level II, sample size is 25.

2.	1/2 TLV (50 ppm)	1 TLV (100 ppm)	2 TLV (200 ppm)	5 TLV (500 ppm)
U	67.5 ppm	125 ppm	250 ppm	625 ppm
L	32.5	75	150	375
$\overline{X}$	54.47	95.97	200.8	498.9
σ	7.52	15.89	23.91	40.0
Q_U	1.73	1.83	2.06	3.15
Q_L	2.92	1.32	2.12	3.10
P_U (Table B-5)	3.84	3.00	1.61	0.018
P_L (Table B-5)	0.063	9.17	1.35	0.024
$P_U + P_L$	3.90	12.17	2.96	0.042
M (Table B-3)	12.57	12.57	12.57	12.57

3. $P_U + P_L$ is less than the maximum allowable error at all concentrations; therefore, the lot is accepted.
4. Least ± percentage tolerance for which the lot would be accepted.

%	26.5	24.7	18.0	12.1
Q_U	1.17	1.81	1.47	1.54
Q_L	2.36	1.30	1.54	1.485
P_U	12.00	3.03	6.74	5.80
P_L	0.518	9.47	5.80	6.53
$P_U + P_L$	12.52	12.50	12.54	12.33
M	12.57	12.57	12.57	12.57

TABLE 11—*Analysis of H_2S detector tubes, Lot 30, using DuPont system.*

	1/2 TLV, 5 ppm	1 TLV, 10 ppm	2 TLV, 20 ppm	5 TLV, 50 ppm
$\overline{X}$	5.42	9.07	20.55	49.80
σ	0.386	0.768	1.644	3.655
CV	0.0712	0.0847	0.0800	0.0733
MCV	0.0775			
b	0.084	−0.093	0.027	−0.003
$/\overline{b}/$	0.004			
OA	±(0.4 + 15.5) = ±15.9			
(OA)c[a]	±(8.4 + 14.2)	±(9.3 + 16.9)	±(2.7 + 16.0)	±(0.3 + 14.7)
	±22.6	±26.2	±18.7	±15.0

[a] (OA)c = estimated accuracy at each individual concentration.

system at an AQL of 6.5% is more lenient than the DuPont system. To make the systems fairly comparable, an AQL of 1.0% in MIL-STD 414 would have to be used. In other words, the DuPont system gives 95% within the tolerance level at 95% confidence level, whereas the MIL-STD method gives 95% at about an 82% confidence level.

Some Approximate Accuracies of Detector Tubes

From production lot data, taken from the last four to eight lots of the detector tubes for the substances listed, the approximate accuracies of the tubes by the DuPont method are given in Table 13.

As can be seen, there is a wide variation of accuracies for the various tubes. With ethylene oxide and formaldehyde, the concentrations are low and reactor tubes must be used, decreasing the accuracy. Carbon tetrachloride is not very reactive and a pyrolyzer is used to break it down with the liberation of chlorine. A chlorine detector tube is used. The amount of breakdown (and chlorine liberated) is dependent on the condition of the pyrolyzer. Most of the tubes, however, fall within the ±25% region.

Errors Using Detector Tubes

Errors are frequently expressed as the relative standard deviation (*RSD*), also known as the coefficient of variance (*CV*).

$$RSD = CV = \frac{\overline{X}}{\sigma}$$

where σ = standard deviation and $\overline{X}$ = average. Two times the *RSD* times 100 gives an estimate in percentage of the precision of the tubes. That is, 95% of the tubes fall within this range at the 95% confidence level. Some causes of error are:

Random (affect precision)

1. Amount of materials in tubes.
2. Packing density of tubes.
3. Variation in readings between observers.
4. Variation in tubes' inside diameter.

TABLE 12—*Analysis of toluene detector tubes, Lot 18, using DuPont system.*

		1/2 TLV, 50 ppm	1 TLV, 100 ppm	2 TLV, 200 ppm	5 TLV, 500 ppm
$\overline{X}$		54.47	95.95	200.8	498.9
σ		7.52	15.89	23.91	40.00
CV		0.138	0.166	0.119	0.080
MCV	0.129				
b		+0.089	−0.0403	+0.004	−0.0022
/b/	0.013				
OA	+(1.3 + 25.8) = ±27.1%				
(OA)c[a]		±(8.9 + 27.6)	±(4.0 + 33.2)	±(0.4 + 23.8)	±(0.2 + 16.0)
		±36.5	±37.2	±24.2	±16.2

[a] (OA)c = estimated accuracy at each individual concentration.

5. Slight variations in temperature, humidity, or pressure.
6. The slope of the calibration curve.

Systematic (affect bias, and perhaps, precision also)

7. Calibration off.
8. Stability during storage.
9. Pump leak.
10. Pump capacity.
11. Using the wrong pump.
12. Interferences.
13. Large temperature, humidity, or pressure variations.

Two items to keep in mind when designing tubes are that (1) the sharper the end point, and (2) the longer the stain for a given challenge concentration (mm/ppm), generally, the more accurate the tube.

There are several means of sharpening the end point. First of all, the finer the indicating chemical, generally, the sharper the line of demarcation between stained and unstained parts will be. Also, the slower the flow rate through the tube, the sharper the end point, since this will allow for more complete reaction between the impregnant and the gas or vapor. More impregnant will also cause more contrast between reacted and unreacted material, making it easier to determine the end point. One final way of improving the end point is to choose an indicator that reacts rapidly with the challenge gas or vapor. This will prevent a diffused-looking end point.

TABLE 13—*Accuracies of various detector tubes.*

Compound	Range	Accuracy	Compound	Range	Accuracy
Ammonia	25 to 250 ppm	±15%	formaldehyde	1 to 5 ppm	±50%
Benzene	5 to 50 ppm	±30%			
Carbon dioxide	0.25 to 2.5%	±12%	hydrogen cyanide	5 to 50 ppm	±15%
Carbon monoxide	25 to 500 ppm	±20%	hydrogen sulfide	5 to 50 ppm	±15%
Carbon tetrachloride	5 to 500 ppm	±70%		50 to 500 ppm	±15%
Chlorine	0.5 to 5 ppm	±20%	nitrogen dioxide	2.5 to 25 ppm	±25%
Ethylene oxide	10 to 50 ppm	±30%	sulfur dioxide	2.5 to 25 ppm	±15%
			toluene	50 to 500 ppm	±25%
			trichloroethylene	50 to 500 ppm	±20%

With regard to the length of stain, there are a number of factors that influence this parameter. Some of these have been touched on in the preceding paragraph, but will be repeated here. A list of some factors affecting stain length are: (1) flow rate, (2) sample volume, (3) reaction rate, (4) completeness of reaction, (5) side reactions, (6) physical adsorption, (7) amount of impregnant, and (8) tube diameter. These are factors that should be considered in the design of a particular detector tube. Of course, challenge concentration, temperature, relative humidity, atmospheric pressure, and interferences will also affect the length of stain, but these are external factors that do not enter directly into design considerations.

From the foregoing discussion, many things that improve the sharpness of the end point in detector tubes tend to decrease the length of stain, and vice versa. For example, a slow flow rate through the tube is advantageous so far as the end point is concerned but reduces the length of stain. Therefore, a balance must be struck between getting the sharpest end point and achieving the longest stain.

Summary

Detector tubes are useful devices for simply, quickly, and inexpensively determining the concentration of many toxic substances. But like most devices, they have limitations regarding applicability. The user must be familiar with these limitations if he is to make proper judgments. Some of the advantages and disadvantages of detector tubes are given here.

Advantages

1. Simple to use.
2. Direct reading: no delay or analytical costs.
3. Low cost per test.
4. Versatile: cover many compounds.
5. Easily portable for field use.
6. Can sample remotely.
7. Good, maintenance-free shelf life.
8. Adequate accuracy for many applications.
9. Long-term tubes measure TWA concentrations.
10. Long-term tubes can be used for personal or area monitoring.
11. Some tubes are certified and, therefore, approved for Occupational Safety and Health Administration (OSHA) use. (Certification program discontinued by NIOSH, but certain tubes still recognized as accepted monitoring methods. The Safety Equipment Institute (SEI) is picking up the certification program.)
12. If no indication on first pump stroke, the same tube can be used again for a different test.

Disadvantages

1. Lack of specificity.
2. Short-term tubes do not give TWA concentrations.
3. Not as accurate as some other, more costly methods.

Because of the many advantageous qualities of detector tubes to indicate toxic substances, they will probably continue to be widely used for this purpose. They certainly give a quick, simple indication if toxic materials are present in the ppm range.

Venkatram Dharmarajan,[1] *Robert D. Lingg,*[1] *Karrol S. Booth,*[1] *and David R. Hackathorn*[1]

Recent Developments in the Sampling and Analysis of Isocyanates in Air

REFERENCE: Dharmarajan, V., Lingg, R. D., Booth, K. S., and Hackathorn, D. R., **"Recent Developments in the Sampling and Analysis of Isocyanates in Air,"** *Sampling and Calibration for Atmospheric Measurements, ASTM STP 957,* J. K. Taylor, Ed., American Society for Testing and Materials, Philadelphia, 1987, pp. 190–202.

ABSTRACT: In the past decade (1975 to 1985) there has been a considerable amount of research activity in the field of sampling and analysis of isocyanates in air. The general trend has been to move away from impinger sampling to solid sorbent or coated filter techniques or both, and the method of analysis has shifted from spectrophotometric to chromatographic techniques.

Dynamic atmospheres of various isocyanates were generated in the laboratory and several sampling and analysis methods for isocyanates were evaluated. The sampling media tested include impingers containing *p*-nitrobenzyl propylamine (nitroreagent) and Marcali solutions, nitroreagent coated glasswool and glassfiber filters coated with nitroreagent, 1-(2-pyridyl)piperazine (PP), and sulfuric acid. The particle size distribution of the 4,4^1 diphenylmethane diisocyanate (MDI) atmosphere was also determined.

The studies show that isocyanates having significant vapor pressure such as toluene diisocyanate (TDI), hexamethylene diisocyanate (HDI), and phenyl isocyanate (PI), are usually present as a vapor in the ambient atmosphere and generally the impinger, the solid sorbent, and the coated filter methods sample them efficiently. However, it has been found in the laboratory generated airborne atmospheres of nonvolatile isocyanates such as MDI and 4,4′ dicyclohexylmethane diisocyanate (Des W or hydrogenated MDI) that the coated filter methods have better collection efficiency than the impingers. Nevertheless, a study from a Swedish team has found quite the opposite results in the field for MDI. In an MDI based polyurethane insulation spraying operation, the impingers collected more than the coated filters in simultaneous sampling.

This paper summarizes our evaluations of the various methods for the sampling and analysis of isocyanates and propose a hypothesis to explain the apparently contradictory finding of the collection efficiency of the filter and impinger method for MDI in the laboratory-generated atmosphere and in the field.

KEY WORDS: isocyanates, diisocyanates, sampling media, analytical methods, industrial hygiene, methods evaluation, sampling, air quality, calibration, atmospheric measurements

Isocyanates, the key ingredient in polyurethane manufacturing, are chemicals containing −NCO functional groups. Mono-, di-, and poly isocyanates are used in industries.

The toxicities and hazards of manufacturing and handling of isocyanates are well recognized [1,2]. Exposure to elevated levels of isocyanates can result in serious respiratory problems and can adversely effect skin and the eyes. Recognizing these potential hazards, the Occupational Safety and Health Administration (OSHA) has promulgated specific ex-

---

[1] Industrial hygiene specialist, industrial hygiene laboratory supervisor, industrial hygiene specialist, and industrial hygiene manager, respectively, Mobay Corporation, Pittsburgh, PA 15205.

posure limits for two isocyanates [3] and the National Institute for Occupational Safety and Health (NIOSH) [1] has a general recommendation of 0.005 ppm time weighted average (TWA) and a 0.02 ppm ceiling exposure limit for all diisocyanates. Among the 700 or so threshold limit values (TLVs) listed in the 1983–84 American Conference of Governmental Industrial Hygienists (ACGIH) TLV booklet, isocyanates came within the first ten chemicals with lowest exposure limits. Because of this unique distinction of isocyanates, a voluminous amount of work has been done in the academia, industries, government, and other institutions in all aspects of isocyanates, such as toxicology, epidemiology, hygiene, and safety.

This paper gives an overview of the general developments in the fields of sampling and analysis of isocyanates in air in the past ten years, supplemented with specific studies done at our laboratories in the past three years. The discussion is restricted to the diisocyanates listed in Table 1.

TABLE 1—*Some commercially important isocyanates and their chemical structures.*

Isocyanate Chemical Name	Acronym	Structure
2,4-Toluene diisocyanate	2,4-TDI	toluene ring with CH$_3$, and two NCO groups (2,4-positions)
2,6-Toluene diisocyanate	2,6-TDI	toluene ring with CH$_3$, and two NCO groups (2,6-positions, OCN— and —NCO)
Methylene diphenyl diisocyanate	MDI	OCN—C$_6$H$_4$—CH$_2$—C$_6$H$_4$—NCO
Phenyl isocyanate	PI	benzene ring with NCO
Hexamethylene diisocyanate	HDI	OCN—(CH$_2$)$_6$—NCO
Biuret of hexamethylene diisocyanate or Desmodur-N	Des N	OCN—(CH$_2$)$_6$—N(H)—C(=O)\>N—(CH$_2$)$_6$—NCO OCN—(CH$_2$)$_6$—N(H)—C(=O)/
Methylene dicyclohexyl diisocyanate (hydrogenated MDI)	H-MDI	OCN—C$_6$H$_{10}$—CH$_2$—C$_6$H$_{10}$—NCO

Significant Developments in Sampling and Analysis of Isocyanates from 1975 to 1985

Prior to the passage of the OSHA Act (1970), most of the isocyanate sampling and analysis were generally restricted to toluene diisocyanate (TDI) and p,p'diphenylmethane diisocyanate (MDI). The most popular sampling and analysis method then was a method referred to as the Marcali Method [4]. The method was specifically developed for TDI, later it was modified to detect MDI [5]. The modified Marcali method was subsequently adopted by NIOSH [6,7] for the determination of TDI and MDI in air. Essentially, the method is based on impinger collection of TDI or MDI in a dilute absorber solution of hydrochloric and acetic acids. The diisocyanates are acid-hydrolyzed to the corresponding amines, which are diazotized with sodium nitrite and coupled to N-1-naphthylethylene diamine to form a colored complex whose optical absorbance is measured with a spectrophotometer. The Marcali method has several limitations such as:

1. The method is not specific; other aromatic isocyanates and amines interfere.
2. The method cannot distinguish between 2,4 and 2,6-TDI, and they have different response factors.
3. The MDI present as aerosol under normal ambient conditions is not sampled efficiently by Marcali impingers.
4. Impinger sampling is not the preferred method for personal sampling.

A similar colorimetric method based on the reagent 1-fluoro-2,4-dinitrobenzene was developed for aliphatic isocyanates by vonEicken [8], which was subsequently improved by Pilz and Johann [9] and by Walker and Pinches [10]. However, this method also had limitations of interferences from aromatic amines and isocyanates.

The specificity, sensitivity, and versatility of airborne isocyanate analysis was significantly improved by a series of high pressure liquid chromatographic (HPLC) methods developed in the past decade. Most of these methods were based on the reaction of isocyanates with amines to form their corresponding urea derivatives, which were separated by HPLC and detected and quantitated using ultraviolet, fluorescence, or electrochemical detectors. These methods were good for both aliphatic and aromatic isocyanates and several isocyanates could be sampled and analyzed simultaneously. They offer potential for analysis of some isocyanate polymers.

Table 2 gives a brief listing of some of the important publications showing the evolution of the recent state of the art in isocyanate analysis. Although significant advances have been made in the development of direct read-out instrumentation for isocyanates, this discussion will be restricted to the classical industrial hygiene sampling and analysis techniques, that is, techniques where sampling and analysis are two separate well-defined functions.

It is clear from Table 2 that, by far, most of the attention has been directed towards HPLC techniques interspersed with some gas chromatographic (GC) methods. The recent GC methods are all based on collection of isocyanates in an acid medium, and derivatizing the hydrolyzed amine using trifluoroacetic, or hepta- or penta-fluorobutyric acid anhydride and quantitating the derivative using GC with electron capture detection. The GC methods cannot distinguish the isocyanates from their corresponding amines and they cannot be used to determine isocyanate polymers.

It is also clear from the table that most of the methods developed depend upon collection of airborne isocyanates in impingers containing a suitable absorber solution and subsequent treatment of the absorber solution in the laboratory, to suit the analytical finish. In industrial hygiene monitoring, generally, impinger collection techniques are recommended for sampling contaminants that are normally present as vapors. Thus, isocyanate impinger sampling

TABLE 2—*Isocyanates sampling and analysis (1975–1985).*

Year	Sampling Medium	Analytical Finish	Isocyanates	Ref
1976	impinger—*p*-nitrobenzyl propyl amine or nitroreagent	HPLC	TDI, MDI, PI, HDI, Des-N	11
1979	impinger—HCl-DMSO[d] 1-fluorodinitrobenzene	UV/VIS[a]	aliphatic HDI, H-MDI	10
1979	impinger—naphthalene-methylamine	HPLC fluorescence	HDI, IPDI	12
1979	impinger—1-(2-pyridyl) piperazine	HPLC	TDI, MDI	13
1979	glassbeads—nitrobenzyl propyl amine	HPLC	TDI, MDI, HDI	14
1980	impinger—HCl/fluoro acids	GC-EC[b]	TDI	15
1980	impinger—methyl amino methyl anthracene	HPLC	TDI, MDI	16
1980	impinger—ethanol	HPLC	TDI, MDI	17
1981	impinger—methyl naphthalene methyl amine	HPLC	TDI, MDI, HDI	18
1981	glassfiber filter—naphthalene methyl amine	HPLC	TDI, MDI, HDI	19
1981	impinger—1-(2-methoxy phenyl) piperazine	HPLC[c] ED	TDI, MDI HDI	20
1982	XAD-2—N-methyl amino methyl anthracene	HPLC	TDI, MDI	21,22
1982	silica gel—diethyl amine	HPLC	MDI	23
1982	glassfiber filter and glasswool—nitro benzyl propyl amine	HPLC	TDI, MDI	24
1982	impinger—HCl/fluoro acid anhydride	GC-EC[b]	HDI	25
1983	impinger—HCl/fluoro acid anhydride	GC-EC[b]	TDI, MDI	26
1983	impinger—alcoholic KOH	HPLC	MDI, TDI, MDA, TDA	27
1983	impinger—1-(2-methoxy phenyl) piperazine	HPLC	total isocyanates	28
1984	impinger—1-(2-methoxy phenyl) piperazine	HPLC[c] ED	total -NCOs	29

[a] UV/VIS = ultraviolet/visible absorption spectrometry.
[b] GC-EC = gas chromatography with electron capture detector.
[c] HPLC-ED = HPLC with electro chemical detection.
[d] DMSO = dimethylsulfoxide.

methods should work well for TDI, HDI, and PI. However, airborne diisocyanates such as MDI, H-MDI, and Des N that have very low vapor pressures at normal ambient temperatures would be present as condensation aerosol (general size range 0.01 to 1 μm). Also in industrial operations where the isocyanates are sprayed under pressure, one would expect a significant amount of the airborne isocyanates to be present as particles, dissolved in or adsorbed to polyol or polyurethane particles and pigments. Under these conditions the impinger sampling techniques may not quantitatively collect all the isocyanates. It is well known that impingers have poor collection efficiency for particles less than about 2 μm in size [30,31]. Recognizing these problems, a few sampling methods based on collection of isocyanates on reagent-coated glassfiber filters, glasswool, or solid sorbents were developed.

Of the several impinger and solid sorbent sampling methods available for isocyanates, the following methods have been tentatively adopted by NIOSH or OSHA, or both.

NIOSH Methods:
1. Nitroreagent in Toluene—impinger method.
2. Nitroreagent coated on glassfiber filter.
3. Nitroreagent coated on glasswool.
4. Marcali impinger method.

OSHA Method:
1. Pyridyl piperazine coated glassfiber filter.

It is reasonable to assume that these methods are the most commonly used methods for the sampling and analysis of isocyanates in United States.

Evaluation of the OSHA and NIOSH Recommended Methods for Isocyanates

In the past three years, we have evaluated the sampling and the analytical aspects of the OSHA and NIOSH recommended methods. The evaluation generally consisted of:

1. Generating an atmosphere of the isocyanate.
2. Sampling the atmosphere concurrently with two or three methods.
3. Analyzing the samples by the respective analytical procedures.
4. Comparing and contrasting the results by different sampling techniques.
5. Determining the relative collection efficiency of different methods and attempting to collect and quantitate the isocyanates lost by the collection medium during sampling.
6. Evaluating analytical parameters such as stability of absorbers on aging, effect of humidity and light, stability of samples after collection, etc.

Isocyanate Atmosphere Generation Techniques

TDI, HDI, and PI

Several different techniques were used for generating isocyanate atmospheres depending upon their chemical and physical properties. Figure 1 shows schematically the generation system used for generating atmospheres of volatile isocyanates. The generation system consists of a motor driven, 100-μL glass syringe, injecting continuously a dilute solution of the isocyanate in toluene, into a small plug of glasswool kept at the end of a 6.35 mm (¼ in.) inside diameter glass tube. Warm air (~60°C) was continuously passed through the glasswool, evaporating the isocyanate uniformly. Injecting into glasswool prevents drop formation and helps to diffuse the isocyanate throughout the glasswool by capillary action, facilitating uniform evaporation. The isocyanate laden air stream was diluted with humidified air and mixed in the mixing chamber and transported to a glass manifold. The manifold had five ports through which air samples could be collected simultaneously into several sampling media. The relative humidity was maintained at 50 to 60%, and was monitored continuously using a Panametrics hygrometer. The sampling sites were at room temperature (~25°C).

Generation of MDI Atmospheres

Dynamic MDI Atmosphere—The continuous injection method could not be used for MDI because of its low volatility. Figure 2 describes the experimental setup used for the generation

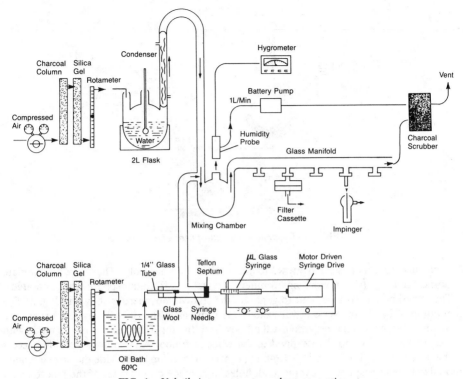

FIG. 1—*Volatile isocyanate atmosphere generation.*

of MDI atmosphere. Neat recrystallized MDI is heated in a 10-mL test tube (Wheaton Micro distillation kit) using a solid aluminum block heater (Pierce Reactitherm). The temperature could be controlled to ±0.5°C. The temperature of MDI was varied from 65 to 115°C to vary the atmosphere concentration. The hot MDI vapor from the test tube was swept off with nitrogen heated to ~60°C, at 5 to 10 L/min flow rate. The MDI vapor was mixed with humid air at room temperature from the humidification system and flowed through a 25.4 by 304.8 mm (1 by 12 in.) long glass tube for mixing, before entering the glass manifold. There were five ports in the manifold for simultaneous sampling of the MDI atmosphere. This MDI atmosphere was referred to as the "dynamic atmosphere" simply to indicate that the lifetime of MDI from the point of generation to the point of sampling was very short, about 4 s. Condensation of MDI was observed in the glass tubing from the point where MDI vapor mixed with the humid air at room temperature. The dilution flow rate was usually three to five times the flow rate of nitrogen, and it cooled the final atmosphere to room temperature or slightly above the room temperature at the point of sampling. The relative humidity of the final atmosphere was between 50 and 60% and the temperature ranged from 25 to 30°C.

Static MDI Atmosphere—Prior to these evaluations, several studies [24,32–34] have indicated that in workplace environment MDI exists predominantly as an aerosol. Attempts to determine the particle counts in the dynamic atmosphere using a portable light scattering SIBATA digital dust counter showed that either there were no MDI particles or the size of the particles was smaller than 0.1 μm, the detection limit of the dust counter. Considering the age of MDI in the dynamic atmosphere, it was inferred that the MDI did not have

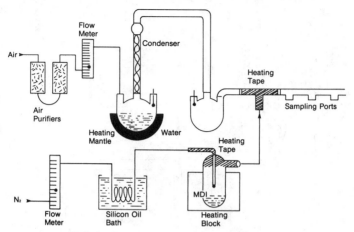

FIG. 2—*Dynamic MDI atmosphere generation.*

enough time to condense and grow, to form detectable aerosols. Therefore to simulate airborne MDI in a typical work environment, a "static" MDI atmosphere was generated.

The experimental arrangement to generate the static atmosphere is shown in Fig. 3. The generation of MDI atmosphere was exactly the same as in the dynamic system except the final atmosphere was charged into a 1.0 m³ box from the top. The cubic box was made with transparent polyethylene sheets glued to the wooden frame. A latch door provided access to the insides of the box. A small 101.6-mm (4-in.) diameter fan kept inside the box circulated and mixed the atmosphere. Several small holes were made on the sides of the box to insert sampling probes. These holes provided the necessary vent to equalize the pressure inside and outside the box. Sampling at different sites inside the box confirmed the uniformity of the MDI atmosphere inside the box. Based on the volume of MDI atmosphere entering the box and assuming perfect mixing, the average age of MDI in the chamber was calculated to be about 45 min. The relative humidity of this atmosphere was between 50 and 60% and the temperature inside the box during the study varied between 25 and 30°C. The box and all the other accessories were kept inside a standing paint spray booth. The SIBATA digital

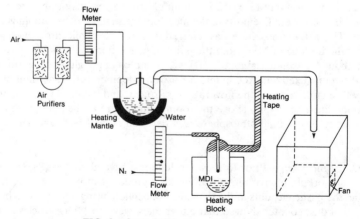

FIG. 3—*Static MDI atmosphere generation.*

dust counter registered counts directly proportional to the concentration of MDI, indicating the presence of MDI aerosol in the system. The particle size distribution of the static MDI atmosphere was determined using a six-stage, parallel-stage impactor [35]. This impactor distributes the particles in six stages as follows: (1) <0.34 μm, (2) <0.75 μm, (3) <1.90 μm, (4) <3.12 μm, (5) <7.34 μm, and (6) total. Based on these particle size measurements, it was concluded that all the MDI aerosol in the static atmosphere was less than 1.9 μm mass median aerodynamic diameter (MMAD); and ~80% of the particles were between 0.34 to 1.9 μm MMAD.

Evaluation of Isocyanate Sampling and Analysis Methods

The following sampling and analytical procedures were compared.
The high pressure liquid chromatographic methods include:

1. Nitroreagent liquid impingers [11].
2. Nitroreagent coated glasswool tubes [36].
3. Nitroreagent coated glassfiber filters [37,24].
4. Pyridyl piperazine coated glassfiber filter [38].
5. Sulfuric acid coated glassfiber method [39].

The spectrophotometric method:

1. Marcali reagent liquid impingers [4,6].

The sulfuric acid coated glassfiber filter method is an unpublished method and therefore the method is briefly described here.

Sulfuric Acid Coated Glassfiber Filter Method

This method was developed by the Stanford Research Institute, under an NIOSH contract, for the sampling and analysis of methylenedianiline (MDA) in air. We adapted the method for MDI to independently check the other sampling methods.

A 37-mm glassfiber filter is soaked with 0.5 mL of 0.27 N H_2SO_4 (sulfuric acid) and dried at 100°C for 1 h. The dried filters loaded in three-piece, 37-mm cassettes with stainless steel backup pads are used for open face sampling. The MDI hydrolyzes in H_2SO_4 to form MDA that reacts with the acid to form the sulfate salt. The filters are desorbed with 4 mL of 0.27 N NaOH (sodium hydroxide) containing 5% acetonitrile. The free MDA is derivatized by adding 1 mL of acetic anhydride and heating for 15 min at 80°C. The derivative is separated using a C_{18}, reverse phase, HPLC column and acetonitrile/water mobile phase. The eluents are measured with an ultraviolet detector at 254 nm wavelength.

The sampling phases of all the preceding methods were evaluated by sampling the isocyanates from the atmospheres. It is important to keep in mind that the isocyanate atmosphere generated was not a standard atmosphere and there was no prior knowledge of its concentration. The methods were evaluated based on their relative performance when they were used simultaneously to sample the same atmosphere. If on comparison one method is found to give significantly lower results, attempts were made to determine the source of losses or explain the source of losses based on current knowledge of aerosol behaviors, or theory of sampling efficiency.

TABLE 3—*Summary of method evaluations.*

Method	Good for	Problems
IMPINGER METHODS		
Marcali	pure 2,4 or 2,6 TDI vapor	poor collection of MDI and PI; interference of aromatic amine and isocyanates. poor collection of aerosols <2 μm.
Nitroreagent	TDI, HDI, PI vapors	poor collection of aerosols <2 μm
SOLID SORBENT METHODS		
Nitroreagent coated glassfiber filter	TDI, MDI, HDI, PI, PI H-MDI aerosol and vapor	light sensitive spurious peaks; blank problems, unstable
PP-coated glassfiber filter	TDI, MDI, HDI PI, aerosol and vapor	with low PP loading; flow rate and humidity affect collection
Sulfuric acid coated glassfiber filter	MDI	amine interference OTI not collected
Nitroreagent coated glasswool	TDI, HDI, PI	light sensitive spurious peaks; blank problems; labor intensive

The following assumptions are implied in this way of evaluating or comparing sampling and analytical methods.

1. The isocyanate atmosphere is homogenous and uniform.
2. There is no bias between different sampling sites or ports.
3. There is no bias in the analytical measurement.
4. There is no bias due to adsorption or deposition in the sampling extension probes, pump calibration, operator errors, or other random variables.
5. If two independent methods, measuring the same atmosphere correlate well within the statistical experimental fluctuations, then both the methods are measuring the true concentration of the atmosphere. Either of the methods can then be considered a standard method and can be used as a reference method to evaluate other methods.

Summary of Isocyanate Sampling and Analysis Method Evaluations

Table 3 summarizes the results of our isocyanate methods evaluation during the past three years.

Impinger Methods

Marcali Method—The Marcali impinger method has been used for a long time for sampling TDI and MDI, and even today it is used by several industries. According to our studies, the use of the Marcali method should be restricted to TDI vapors and then, when only one isomer is present. If both 2,4 and 2,6-TDI isomers are present in the environment as is the case in most of the applications, the Marcali method cannot be used because of different response factors by the isomers. Rando et al. [40] has shown that the isomeric composition of TDI in air can vary from 0 to 100% for an isomer, independent of the composition of the starting material. However, recently in 1985, Rando and Hammad [41] have reported a modified Marcali method for the determination of total TDI when both 2,4 and 2,6 isomers are present.

For a long time, MDI has also been sampled and analyzed by the Marcali method, implicitly assuming airborne MDI to be vapor. However, several recent studies [24,32–34] have shown airborne MDI to be predominantly aerosol and MDI collection losses of 50% using the Marcali method. Other developments in impinger collection of MDI are discussed later. We also found that Marcali impingers do not collect phenyl isocyanate (PI) efficiently, although airborne PI exists as vapor at normal temperature and pressure. This may be due to un-favorable reaction kinetics of the hydrolysis of PI.

Nitroreagent Method—The nitroreagent impinger method was tested for TDI, HDI, and PI vapors and was found to be acceptable. There are no known interferences; however, one should be cautioned that if the toluene solvent is not completely evaporated off in the sample treatment step, it elutes as a peak and interferes with the 2,6-TDI peak. The nitroreagent absorber solution should be used within two weeks after preparation, and care should be taken to protect it from sun and fluorescent light. If exposed to strong light, nitroreagent decomposes to give spurious peaks in the chromatogram that can interfere with TDI or MDI peaks.

Although nitroreagent impinger is used for sampling MDI, Des N, and H-MDI, there could be collection problems depending upon the particle size distribution of these iso-cyanates in air. Using laboratory-generated static MDI atmosphere, we have conclusively shown that standard, matched midget impinger (30 mL, jet bore = 1mm, and 0.5-cm impaction distance) operated at 1 L/min flow rate can show losses ranging from 40 to 78% at MDI concentrations of 0.022 to 0.07 ppm determined by sulfuric acid and nitroreagent coated filter methods [34]. The MDI, escaping the impinger, were collected on back-up sulfuric acid filters and quantitated. All the MDI particles in the static atmosphere were less than 2 μm and the poor collection of particles in this size range by impingers are well known [30,31].

However, a recent study by Andersson et al. [21] showed that, in a field spraying operation, the impingers collected more MDI than nitroreagent coated glassfiber filter method in side-by-side sampling. In the same paper, the authors generated MDI atmosphere in the labo-ratory using a Collison nebulizer and found that the coated filters collected more MDI than the impingers concurring with our observations. The authors tried to explain this anomaly between the laboratory-generated and field results by proposing some complex MDI-specie in the field operation that dissolved in the impinger solution to form MDI.

Gudehn [42] in a later publication provided evidence for the formation of MDI-hydrate using various techniques such as nuclear magnetic resonance (NMR), infrared (IR), mass spectrometry, and differential scanning colorimetry. The author theorized that the MDI-hydrate dissolves in the impinger solvent, liberating MDI, thus accounting for higher MDI in impingers. This does not occur in coated filters.

However, their results could also be explained based on particle size distribution and sampling characteristics. The authors did not report the particle size distribution of the spray in the field. Suppose the mass median aerodynamic diameter (MMAD) of the spray was 15 μm, and an impinger would collect this atmosphere with 99 + % efficiency. Dharmarajan [33] has found a particle size of 11 μm MMAD in an MDI spray operation. The reagent coated filters will also stop the particles with 100% efficiency, but because of the large size of the particle, only the MDI in intimate contact with the reagent on the surface of the filters react. Therefore, the MDI inside and on the outer surface of the aerosol will not react with the coated reagent but will be consumed by polyols or other reactive species used in the operation. Thus, it appears that in a spraying operation a combination of an impinger followed with a coated filter may be needed to quantitatively collect MDI. More testing and research is needed to definitively explain the differences in collection efficiencies of impingers and coated filters in spray operations.

Reagent Coated Solid Sorbents

The nitroreagent and the 1-(2-pyridyl)piperazine (1 mg PP on 13-mm filters) coated glassfiber filters were tested for TDI, MDI, HDI, and PI in laboratory-generated static and dynamic atmospheres. Both filter methods collected all the isocyanates efficiently and the analysis gave equivalent concentrations. The static MDI atmospheres were tested using sulfuric acid, nitroreagent, and 1-(2-pyridyl)piperazine (1 mg of PP on 13-mm filter) coated filter methods and all were found to give equivalent results. However, if only 0.1-mg PP is coated on 13-mm filter, the filter's collection efficiency for all isocyanates was poor and also the collection efficiency increased on decreasing the flow rate from 2 to 0.5 L/min. This shows that when 13-mm-diameter filters coated with 0.1-mg PP are used, there is a flow rate effect on collection efficiency. When the same experiments were performed with 37-mm-diameter filter with 0.1-mg PP (as recommended by OSHA), TDI, HDI, and PI were collected efficiently and there was no flow rate effect. These experiments clearly show that differences in the collection efficiencies are due to the differences in the face velocities of 13-mm and 37-mm-diameter filters. The face velocity of the 13-mm-diameter filter (1.5 m/min for 2 L/min flow) is eight times greater than the face velocity of the 37-mm filter. It is interesting to note that although the density of the PP reagent on the 37-mm filter is eight times less than that on 13-mm filter, at the same flow rate, there is still efficient collection in 37-mm filter, showing that the collection clearly depends on kinetics of the reaction.

When OSHA used 0.1-mg PP on 37-mm filters, they showed the collection efficiency of TDI was affected by high humidity. Based on this observation, they recommended a maximum 15-L sample volume for TDI. However, by using 1.0-mg PP on 13 mm filter, we found no humidity effects for up to 4 h or 200 L of 80% relative humidity air sample. As of this writing, the studies are still in progress, but it is safe to conclude that increasing the amount of PP on the 37-mm filter from 0.1 mg to 1.0 mg will improve the collection efficiency and minimize the humidity effects.

The nitroreagent filters were adversely affected on exposure to sunlight or even fluorescent light. Spurious peaks in the chromatogram due to degradation products of the nitroreagent interfered with isocyanate analysis. The PP filters were not affected by exposure to light. Both PP and nitroreagent filters (kept in dark) were stable for two weeks before and after collection of samples. The PP-filter method is the preferred method for isocyanates because of its insensitivity to light and the ease of preparation of the filters.

The nitroreagent coated glasswool was found adequate for vaporous isocyanates (TDI, HDI, and PI). However, because of the difficulties involved in making nitroreagent coated glasswool tubes and also the nitroreagent sensitivity to light, this method is not recommended for routine use.

Sulfuric acid coated glassfiber filters were tested for MDI. The method was found to be equivalent to nitroreagent or PP filter method. However, a limitation of the method is the interference due to MDA. The method cannot distinguish between MDI and MDA. Surprisingly, ortho tolyl isocyanate vapor went right through the sulfuric acid coated filter with 0% collection. We suspect sulfuric acid filters may also have poor collection efficiency for phenyl isocyanate because Marcali impinger (3.5% hydrochloric acid and 2.3% acetic acid) do not collect them efficiently. We have not tested sulfuric acid filters for other isocyanates.

Conclusions

Sampling and analysis of airborne isocyanates is a challenging problem. In the past decade alone, more than 100 papers have been published dealing with this subject and significant advances have been made in both the sampling and analysis phases. As a result, the versatility, reliability, and sensitivity with which one can characterize isocyanates in work environments was greatly enhanced. However, the areas that we feel need more research include:

1. Resolution of the correct sampling technique needed to characterize isocyanate in an aerosol form or when sprayed.
2. Determination of the particle size distribution of a typical isocyanate spray operation in the field and determining the isocyanate concentration in the various particle size ranges.
3. Determination of the isocyanate in the vapor and aerosol form in a spray system.
4. Effect of different matrices on the analysis of isocyanates.
5. Development of methods to determine polymeric isocyanates in air.

References

[1] Criteria for A Recommended Standard, "Occupational Exposure to Diisocyanates," National Institute for Occupational Safety and Health, U. S. Department of Health, Education, and Welfare, 1978.
[2] *Documentation of the Threshold Limit Values,* 4th Ed., American Conference of Governmental Industrial Hygienists, Inc., 1984, pp. 403.1–403.6.
[3] General Industry Occupational Safety and Health Standards (29CFR1910), U. S. Dept. of Labor, OSHA 2206, revised, 1981.
[4] Marcali, K., *Analytical Chemistry,* Vol. 29, 1957, pp. 552–558.
[5] Grim, K. E. and Linch, A. L., *Journal,* American Industrial Hygiene Association, Vol. 25, 1964, pp. 285–290.
[6] *NIOSH Manual of Analytical Methods,* 2nd Ed., Vol. 1, P & CAM No. 141, "2,4-TDI in Air," and No. 142, "MDI in Air," U. S. Department of Health, Education, and Welfare, Cincinnati, 1977.
[7] Criteria for a Recommended Standard—Occupational Exposure to Toluene Diisocyanate, U. S. Department of Health, Education, and Welfare, National Institute for Occupational Safety and Health, 1973.
[8] vonEicken, S., *Mikrochimica Acta,* Vol. 6, 1958, p. 731.
[9] Pilz, W. and Johann, I., *Microchimica Acta,* Vol. 27, 1970, pp. 351–358.
[10] Walker, R. F. and Pinches, M. A., *Analyst,* Vol. 104, 1979, pp. 928–936.
[11] Dunlop, K. L., Sandridge, R. L., and Keller, J., *Analytical Chemistry,* Vol. 48, 1976, pp. 497–499.
[12] Levine, S. P., Hoggar, J. H., Chaldek, E., Jungelaws, G., and Gerlock, J. L., *Analytical Chemistry,* Vol. 51, 1979, pp. 1106–1109.
[13] Hardy, H. L. and Walker, R. F., *Analyst* (London), Vol. 104, 1979, pp. 890–891.
[14] Keller, J. and Sandridge, R. L., *Analytical Chemistry,* Vol. 51, 1979, pp. 1868–1870.
[15] Ebell, G. F., Flemming, D. E., Genovese, J. H., and Taylor, G. A., *Annals of Occupational Hygiene,* Vol. 23, 1980, pp. 185–188.
[16] Sango, C. and Zimerson, E., *Journal of Liquid Chromatography,* Vol. 3, 1980, pp. 971–990.
[17] Bagon, D. A. and Purnell, C. J., *Journal of Chromatography,* Vol. 190, 1980, pp. 175–182.
[18] Kormos, L. H., Sandridge, R. L., and Keller, J., *Analytical Chemistry,* Vol. 53, 1981, pp. 1122–1125.
[19] Rappoport, S. M., "Development of an Air Sampling and Analytical Method for 1,6 Hexamethylene Diisocyanate," National Technical Information Service, Report No. AD-A 105002, 1981.

[20] Warwick, C. J., Bagon, D. A., and Purnell, C. J., *Analyst,* Vol. 106, 1981, pp. 676–685.
[21] Andersson, K., Gudehn, A., Levin, J. O. and Nilsson, C. A., *Journal,* American Industrial Hygiene Association, Vol. 44, 1983, pp. 802–808.
[22] Andersson, K., Gudehn, A., Levin, J. O., and Nilsson, C. A., *Chemosphere,* Vol. 11, 1982, pp. 3–10.
[23] Lipski, K., *Annals of Occupational Hygiene,* Vol. 25, 1982, pp. 1–4.
[24] Tucker, S. P. and Arnold, J. E., *Analytical Chemistry,* Vol. 54, 1982, pp. 1137–1141.
[25] Esposito, G. G. and Dolzine, T. W., *Analytical Chemistry,* Vol. 54, 1982, pp. 1572–1575.
[26] Bishop, R. W., Ayers, T. A., and Esposito, G. G., *Journal,* American Industrial Hygiene Association, Vol. 44, 1983, pp. 151–155.
[27] Nieminen, E. H., Saarinen L. H., and Lasko, J. T., *Journal of Liquid Chromatography,* Vol. 6, 1983, pp. 453–469.
[28] *NIOSH Manual of Analytical Methods,* 3rd Ed., Peter M. Eller, Ed., Vol. 2, Method 5505, National Institute for Occupational Safety and Health, U. S. Department of Health, Education, and Welfare, 1984.
[29] Bagon, D. A., Warwick, C. J., and Brown, R. H., *Journal,* American Industrial Hygiene Association, Vol. 45, 1984, pp. 39–43.
[30] Murphy, C. H., *Handbook of Particle Sampling and Analysis Methods,* Verlag Chemie International, 1984, p. 184.
[31] Cadle, R. D., *The Measurement of Airborne Particles,* Wiley, New York, 1975, p. 114.
[32] Dharmarajan, V. and Weill, H., *Journal,* American Industrial Hygiene Association, Vol. 39, 1978, pp. 737–744.
[33] Dharmarajan, V., *Journal of Environmental Pathology and Toxicology,* Vol. 2, 1979, pp. 1–8.
[34] Booth, K. S., Dharmarajan, V., Lingg, R. D., and Darr, W. C., *Proceedings,* SPI 28th Annual Conference, Polyurethane—Marketing and Technology—Partners in Progress, Society of the Plastics Industry, 1984.
[35] Tuchman, D. P., Esman, N. A., and Weyel, D. A., *Journal,* American Industrial Hygiene Association, Vol. 47, 1986, pp. 55–58.
[36] *NIOSH Manual of Analytical Methods,* Vol. 6, P & CAM No. 326, U. S. Department of Health, Education, and Welfare, Cincinnati, 1980.
[37] *NIOSH Manual of Analytical Methods,* Vol. 7, P & CAM No. 347, U. S. Department of Health, Education, and Welfare, Cincinnati, 1981.
[38] OSHA Analytical Laboratory, Method No. 42 and 47, Occupational Safety and Health Administration, Salt Lake City, 1983.
[39] Draft Method, "Description of Preliminary Method for Sampling and Analysis of Aromatic Diamines," developed by Stanford Research Institute under contract from NIOSH, 1984, private communication.
[40] Rando, R. J., Abdel-Kader, H. M., and Hammad, Y. Y., *Journal,* American Industrial Hygiene Association, Vol. 45, 1984, pp. 199–203.
[41] Rando, R. J. and Hammad, Y. Y., *Journal,* American Industrial Hygiene Association, Vol. 46, 1985, pp. 206–210.
[42] Gude'hn, A., *Journal,* American Industrial Hygiene Association, Vol. 46, 1985, pp. 142–149.

George E. Podolak,[1] *Richard A. Cassidy,*[1] *George G. Esposito,*[1]
and Donald J. Kippenberger[1]

Collection and Analysis of Airborne Hexamethylene Diisocyanate by a Modified OSHA Method

REFERENCE: Podolak, G. E., Cassidy, R. A., Esposito, G. G., and Kippenberger, D. J.,
**"Collection and Analysis of Airborne Hexamethylene Diisocyanate by a Modified OSHA
Method,"** *Sampling and Calibration for Atmospheric Measurements, ASTM STP 957,* J. K.
Taylor, Ed., American Society for Testing and Materials, Philadelphia, 1987, pp. 203–212.

ABSTRACT: A modified Occupational Safety and Health Administration (OSHA) method
for the determination of 1,6-hexamethylene diisocyanate (HDI) in air was tested in the laboratory and applied in the analysis of spray paint environments. Field tests included side-by-side comparison of alternative procedures of HDI sampling, along with tests to define the
stability of 1-(2-pyridyl) piperazine [1-(2-P)P] impregnated filters, and the effect of increased
reagent loading. Since the method employs 1-(2-P)P in excess to form the HDI-urea derivative
that is then analyzed by high-performance liquid chromatography (HPLC), there is unreacted
reagent present in samples. This residual (unreacted) portion causes HPLC interferences,
extended analysis time, and deterioration of HPLC columns. To avoid these undesirable
characteristics, a means for the removal of "excess" reagent was developed. This paper describes the application of a multi-port valve and short precolumns for facilitating the removal
of "excess" reagent and presents data generated by this improved method.

KEY WORDS: air quality, calibration, sampling, atmospheric measurements, 1-(2-pyridyl)
piperazine, 1,6-hexamethylene diisocyanate (HDI), airborne HDI, polyurethane, glass-fiber
filter, impinger, adsorbent tube, filter cassette, impregnated filter, sampler, Marcali solution,
high-performance liquid chromatography, reverse phase, column-switching valve, flow rate,
air velocity, precolumns

Currently, a polyurethane material with superior resistance to chemicals and ultraviolet
radiation is used by the U. S. Army in lieu of other paints for coating ordnance equipment.
This Chemical Agent Resistant Coating (CARC) is derived from 1,6-hexamethylene diisocyanate (HDI) and contains highly reactive isocyanate compounds. Since diisocyanates are
known to be hazardous to human life, the atmospheres around CARC paint spray operations
need to be monitored.

At the present time, the best option for the analysis of airborne HDI is based on the
application of high-performance liquid chromatographic (HPLC) techniques. The HPLC
approach incorporates *in situ* derivatization of isocyanates to stable ureas during sample
collection followed by HPLC separation, and detection by ultraviolet or fluorescence spectroscopy. Conversions of isocyanates to ureas have been commonly accomplished with N-
4-nitrobenzyl-N-N-propylamine (nitro reagent) [1–4]; however, because of its decomposition
in oxidizing or reducing atmospheres [5] or both, the use of nitro reagent has been replaced
by other urea forming reagents. Recently, the use of N-methyl-1-naphthalene methylamine,

[1] U. S. Army Environmental Hygiene Agency, Aberdeen Proving Ground, MD 21010-5422.

9-(N-methylamine-ether) anthracene, and 1-(2-pyridyl) piperazine reagents has been described in published HPLC isocyanate procedures [6,7]. Of these, 1-(2-pyridyl) piperazine [1-(2-P)P] has become the preferred derivatization compound [8–10].

The Occupational Safety and Health Administration's (OSHA) HDI method that was part of this study uses glass-fiber filters impregnated with 1-(2-P)P for sample collection and HPLC with a C_8 reverse-phase column for analysis [10]. A common difficulty with HDI derivatization reagents is that the excess reagent (unreacted) present in samples greatly complicates the HPLC analysis. It often precludes simpler isocratic analysis, mandating gradient elution and strict adherence to specific columns. Furthermore, the excess reagent causes rapid deterioration of costly HPLC columns, requiring their frequent replacement [5,11]. To complicate matters, it was deemed necessary to increase the amount of 1-(2-P)P used by OSHA ten-fold to assure total conversion of the diisocyanate when sampling relatively high HDI atmospheres.

For these reasons, the first phase of this study was concerned with a means for removal of unreacted reagent. The second phase of the study encompassed three field trips and was directed toward determining the effectiveness of the modified method when compared to the OSHA procedure. At the same time, other tests were conducted to determine the stability of samples during transit and storage, and to compare the collection efficiencies of tubes, impingers, and filters.

Experimental

Apparatus

A Waters Associates HPLC system equipped with a Model 440 ultraviolet detector set at 254 nm, a Model U6K injector, two Model 6000A solvent delivery pumps, a Model 660 solvent programmer, a Radial Compression Module, and 10-μm reverse phase C_8 and C_{18} Radial Pak cartridge columns were used. In addition, 5-μm C_8 and C_{18} Merck Co. glass Kartushe's (Lichrosorb, inside diameter 3 mm, length 150 mm) were utilized, either singularly or two of one kind in tandem. Plumbed into the system was a Valco Instruments, Inc. ten-port electrically-actuated column switching valve that was preceded by two Perkin-Elmer 35-mm precolumns with 3-μm C_{18} packing. The signal from the system's ultraviolet detector was plotted on a HP 3390A integrator. This unit was equipped with an automatic start-up device interconnected with the U6K injector.

Method

The OSHA sampling and analytical procedures used in this study are described in detail by Burright [10]. Briefly, OSHA's method stipulates the collection of samples on a 37-mm glass fiber filter, coated with 0.1 mg of 1-(2-P)P. At least 15 L of air is drawn at a flow rate of 1 L/min. The HDI is converted to urea derivative by reaction with 1-(2-P)P. The derivative is extracted from filters with 90/10 [volume/volume (V/V)] acetonitrile/dimethyl sulfoxide (DMSO) and analyzed by reverse-phase HPLC with the use of 25 cm by 4.7 mm inside diameter, stainless steel, C_8 column. Mobile phase of 37.5/62.5 acetonitrile/water (V/V)— 0.01 M ammonium acetate buffered to pH 6.2 with acetic acid is used. Flow rate of 1 mL/min is applied. Peak quantitation (ultraviolet, 254 nm) is achieved by comparison with standards and reported in terms of diisocyanate rather than the urea derivative. The present study modifies the OSHA method to include an increased amount of reagent on filters and a mechanical procedure for removing excess 1-(2-P)P reagent from samples with a multiport valve. The schematic of the ancillary equipment established to facilitate removal of 1-(2-

P)P is shown in Fig. 1. Instrumental conditions differ somewhat too. The column is a Waters 8-μm C_{18} Radial Pak cartridge. Two Perkin-Elmer precolumns are required. They are 35-mm long and contain 3-μm C_{18} packing. Mobile phase is the same as OSHA's. Two pumps are utilized. Flow rates are set at 0.3 mL/min for Pump No. 1 and 0.8 mL/min for Pump No. 2. In case of alternate conditions for use with two Merck C_{18} columns (in tandem), flow rates of 0.5 mL/min (Pump No. 1) and 0.4 mL/min (Pump No. 2) are applicable, while, other parameters remain the same (Fig. 1).

Laboratory Phase

The urea derivative of HDI was prepared by reacting HDI in methylene chloride with 1-(2-P)P (using heat) and precipitating it with hexane. The product was purified by double recrystallization from methylene chloride with hexane. It was then dried and placed in a tightly sealed amber glass vial in a dessicator that was kept under normal refrigeration. Working standards were made to contain 0.5, 2.5, and 5.0 ng of HDI equivalent per μL in 90:10 (V/V) acetonitrile/dimethyl sulfoxide (ACN/DMSO). These standards were spiked with varying amounts of 1-(2-P)P to test the effectiveness of the system for removal of excess reagent.

To establish the shelf life of working standards, aliquots of prepared standards were withdrawn periodically and analyzed during the course of six months. All standard solutions were kept in dark bottles and sealed with Teflon coated septa. They were left on the bench during working hours on days when used, otherwise they were stored in a refrigerator.

To determine the efficiency of the method, two HDI spiking solutions were prepared in methylene chloride to contain 10 and 100 ng of HDI per microlitre. Appropriate amounts of spikes were introduced directly on the 1-(2-P)P treated filters held in separate 4-mL Wheaton vials. The filters (2-μm glass fiber, 37 mm, Millipore, Inc.) contained 1 mg of 1-(2-P)P. Two millilitres of ACN/DMSO extracting solution were added to the vials; they were sealed, shaken, and subsequently analyzed by HPLC. Altogether six HDI "spikes" were prepared for seven concentration levels tested (1 to 25 ppb).

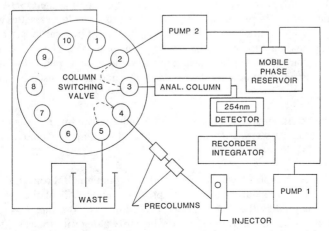

FIG. 1—*HPLC system for analysis of CARC samples (schematic).*

Field Validation

Field tests were conducted to study various parameters of sample collection, storage, and analysis. On-site multiple side-by-side area sampling was conducted along with limited field analysis. This approach served to extend the validation data base to HDI atmospheres that are inherently difficult to generate in the laboratory because of their high reactivity.

In the first field survey, 16 sets of HDI samplers were exposed in a paint spray booth during the painting of large Army trailers with CARC paint. Each test set was arranged in a circle and consisted of the following samplers:

1. Three impingers each with 0.1 mg of 1-(2-P)P in 10 mL of toluene.
2. Three diatomaceous earth (100 mg) tubes impregnated with 1-(2-P)P (0.2 mg).
3. Three filter cassettes containing glass-fiber filters (37 mm) impregnated with 0.1 mg 1-(2-P)P. Preparation of these samplers was done according to the OSHA method [*10*].
4. One impinger containing 10 mL of Marcali solution [*11*].

From each of the three types of 1-(2-P)P sets, one sampler was analyzed on-site, another was refrigerated immediately and shipped to the laboratory on dry ice, and the remaining one was kept at ambient temperature until it was analyzed in the laboratory. In turn, the Marcali solutions were shipped and stored at ambient temperature until analyzed. The total elapsed time before analysis did not exceed 21 days for any of the samples. All 1-(2-P)P-based samples were analyzed in accordance with OSHA's HPLC procedure [*10*]. Marcali solution-based samples were analyzed by gas chromatography (GC).

The second field study dealt primarily with the aspects of HDI sampling on filters. For that purpose, three different batches of filters were prepared. Batch 1 filters were impregnated each with 1.0 mg of 1-(2-P)P; Batch 2 filters contained 2.0 mg; and Batch 3 contained 4.0 mg of 1-(2-P)P per filter. The impregnated filters were then placed in plastic cassettes fitted with back-up pads [*10*]. Additionally, Batch 2 filters had one half and, Batch 3 filters had three quarters of their surfaces blocked with very precise "cut-outs" that were made from a dense paper foil. Inasmuch as only the uncovered parts of filters were used, the amount of 1-(2-P)P in these samplers (Batches 2 and 3) was equivalent to the one held by Batch 1 filters (1.0 mg). Thus, the surface of the samplers was decreased to one half (Batch 2) and one fourth (Batch 3) of the normal size filter (Batch 1). These filters were tested in a series of nine; three samplers from each batch.

The third field study was devised for the purpose of comparing the OSHA method with the modified method developed in this laboratory. For this test, filter-cassettes were mounted in 3.175 mm (⅛ in.) pressed fiberboard that was held upright by a cardboard carton. Pumps were encased in the carton to protect them from paint spray mist. The filter-cassette holders were placed in a paint spray booth adjacent to the vehicles being painted with CARC. The following series of samplers denoted by letters A, B, and C were used in each of the test sets:

A = filter cassettes containing glass-fiber filters impregnated with 0.1 mg of 1-(2-P)P.
B and C = filter cassettes containing glass-fiber filters impregnated with 1 mg of 1-(2-P)P.

The samplers in each sample set were randomized to account for HDI concentration gradients that could have occurred during the paint spray operation. All of the A samples were processed and analyzed by the OSHA HPLC procedure. The B samples were desorbed in the field, sealed in vials, and returned to the laboratory along with the cassettes containing the C samples. Both the B and C samples were analyzed using the modified HPLC procedure. The rationale for shipping the B samples in solution was to determine the stability of the samples. (A prior in-house study proved that the solubilized samples remain stable for over one month at ambient temperature). In addition, nine impingers containing 10 mL of Marcali solution were used. The collected samples were analyzed by gas chromatography [11,12]. This test was conducted because of an anomaly of high HDI results recurring with the use of impinger procedure in the first phase of field sampling.

Results

Laboratory Phase

In the past to prevent the problems caused by the presence of unreacted derivatization reagent in samples, residual reagent was reacted with p-tolyl isocyanate [13] or acetic anhydride [14] to prevent tailing and column deterioration. Since these treatments introduce other chemicals to the HPLC system, it was decided that a better approach would be to mechanically remove 1-(2-P)P from the injected sample prior to its reaching the analytical column. The schematic of the HPLC system including a ten-port valve and two specified short precolumns facilitating such removal of 1-(2-P)P is shown in Fig. 1. In this configuration, the 1 mg of unreacted 1-(2-P)P per sample had no effect on the analysis; it was totally, and easily, removed from the 25 µL sample aliquot. For comparison, Fig. 2 shows two chromatograms presented side-by-side, one with the 1-(2-P)P flushed to waste and the other with the reagent peak emerging late (approximately 60 min), thus demonstrating the effectiveness of the new method to greatly shorten the analysis time. This technique also proved valuable for the analysis of field samples by disposing of other compounds that elute after HDI, for example, diisocyanate oligomers.

The improved analytical procedure was then applied to establish the shelf life of the working standards. These results are shown in Table 1 and reveal that the standards can be safely stored up to six months. Next, the efficiency of the method using spiked filters was tested. The percent mean recoveries and relative standard deviations from this test of six spikes each at seven concentration levels are presented in Table 2. Recoveries are consistent

TABLE 1—Results of analysis[a] of working standards (in counts/ng) within a span of six months.

Sample No.	Analysis after Preparation, Date	Average Counts/ng, x	Standard Deviation Counts	Coefficient of Variation (Cv), x
1	1st day	57 878	3 593	6.21
2	4th day	58 916	4 324	7.34
3	10th day	56 957	4 317	7.58
4	3 weeks	61 627	7 605	12.34
5	5 weeks	62 525	3 626	5.80
6	9 weeks	60 463	5 526	9.14
7	12 weeks	52 836	7 450	14.10
8	4 months	59 285	3 047	5.14
9	4.5 months	51 386	3 813	7.42
10	6 months	56 406	3 452	6.12
	Average	57 828	4 675	8.12

[a] Results represent average from six liquid chromatographic runs, obtained by injecting random sizes (25 µL or less) of each of the three standards (0.5, 2.5, and 5.0 ng/µg HDI) twice.

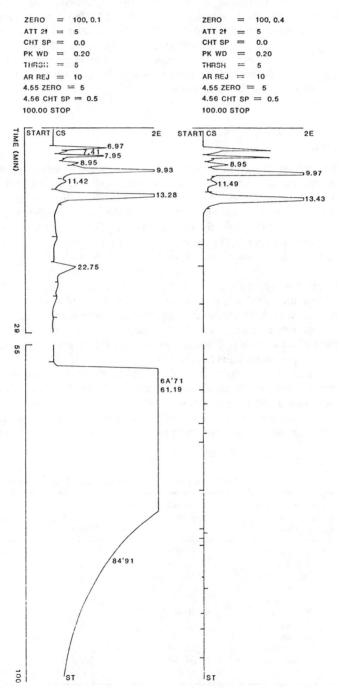

FIG. 2—*Comparison of chromatograms from OSHA method* (left) *and modified OSHA method* (right).

TABLE 2—*HDI recoveries from laboratory prepared samples (averages of six "spikes" at seven concentration levels).*

Theoretical Value, ppb[a]	Found Value, ppb[a]	% Recovery, x	Relative Standard Deviation, %
25	21.67	86.68	2.88
20	20.33	101.65	13.94
10	9.73	97.30	1.61
5	5.12	102.47	15.86
3	2.88	95.89	5.58
2	1.86	93.00	2.85
1	1.12	112.00	8.79
Average	. . .	98.43	7.36

[a] ppb = values calculated on basis of 60 L (volume air) per sample.

with theoretical values for HDI in the range of 1 to 25 ppb (calculated on basis of a 60-L air sample).

Field Phase

The primary objective of the first field study was to obtain a comparison of impinger and nonimpinger HDI sampling methods. Furthermore, tests were conducted to provide insight regarding the stability of the urea derivatives during transit and storage. A total of 160 samples were collected at a CARC spray facility using the three different modes of collection: impingers, filters, and tubes. Some samples were analyzed at the test site while others were sent under varying conditions to the laboratory. No discernable differences were noted between results of the samples analyzed immediately at the site and those that were shipped to the laboratory for analysis. Consequently, average results for each collection mode set-up were calculated regardless of post-test treatment of the samples. Data comparing the sampling methods are presented in Table 3. No data for impinger samples collected in pyridyl piperazine/toluene sorbent are included. Data reliability from this procedure was poor because of acute evaporation losses and breakage and spillage of samples that occurred in transit. As shown in Table 3, the GC impinger method produced consistently higher results when compared with HPLC methods. These high results may be attributed to the presence of an interfering compound, such as hexamethylene diamine. Conversely, competitive reactions with HDI may take place between the reagent and other paint components that would suppress the results from reagent-impregnated filters [15]. The tubes gave the lowest results. Furthermore, open-face filters provided higher results than closed-face filters. The low results for the tubes and closed-face cassettes can be explained by the fact that in each case all of the sample impinged on a small area of collection medium thereby consuming all of the reagent available at a specific site, thus producing low results. Subsequently, impregnated tubes and closed-face filter samplers were deleted from the sampling scheme.

The second field study was conducted using filters impregnated with a ten times greater amount of 1-(2-P)P than called for in the OSHA procedure. This reagent increase was decided upon to ensure the total *in situ* derivatization of large amounts of HDI when encountered in sampled atmospheres. Relatively high HDI concentrations are generally found in CARC spraying operations of large items (for example, trucks). At the same time an assessment was made of the recoveries of HDI collection in relation to the increased air velocities through the 1-(2-P)P impregnated filters. The filters for this purpose were prepared in three different configurations and their 1-(2-P)P content adjusted accordingly (see Ex-

TABLE 3—*Results from first field survey (airborne HDI concentrations measured at a CARC application site by different sampling and instrumental procedures—each entry average of three samples).*[a]

Test No.	GC[b] Impinger, ppb	HPLC[c] Filters, ppb	HPLC[c] Tubes, ppb
1	6.53	6.04	2.11
2	6.57	4.93	1.97
3	8.38	7.09	4.13
4	10.52	3.62	3.99
5	11.97	4.74	2.90
6	7.28	2.47	3.17
7	21.41	7.46	0.86
8	17.34	6.37	2.80
9	21.58	5.74	2.38
10	18.67	5.73	5.05
11	32.00	3.50	5.78
12	28.10	12.87	5.57
13	11.76	4.97	3.12
14	12.20	4.60	2.30
15	13.20	1.30	4.30
16	18.20	2.42	4.13

[a] Samples 1 to 3 were collected in open-face cassettes, the rest were collected in the closed-face mode.
[b] Marcali solution (10 mL).
[c] 1-(2-P)P-based samplers.

TABLE 4—*Results from second field test (airborne HDI in ppm—samples collected using three sizes of filters with adjusted 1-(2-P)P loadings).*

Test Set No.	Sample No.	Filter Surface Area 1	Filter Surface Area ½	Filter Surface Area ¼
1	1	4.3	1.9	2.3
	1[a]	nd[b]	nd	nd
	2	9.0	3.5	0.3
	2[a]	nd	nd	nd
	3	5.5	2.4	0.6
	3[a]	nd	nd	nd
2	4	191.4	59.4	15.6
	4[a]	nd	nd	1.8
	5	179.2	85.0	33.2
	5[a]	nd	nd	1.2
	6	222.8	44.2	88.4
	6[a]	nd	nd	0.7
3	7	2.1	1.3	1.1
	7[a]	nd	nd	nd
	8	2.6	1.7	1.2
	8[a]	nd	nd	nd
	9	4.2	1.4	0.5
	9[a]	nd	nd	nd

[a] Back-up filters.
[b] nd = none detected.

perimental section). To determine a potential HDI breakthrough at increased air velocities, all samplers were backed up with secondary filters irrespective of the configuration of the employed filter. The cassettes holding back-up filters were placed downstream of each test filter. The results from the test are entered in Table 4. Note, the results from this test are given in ppm. They are grouped in test sets and aligned according to the size of filters. Filters within each set were assigned positions at random. The sample results from Set 2 exhibit relatively high HDI concentrations. They were obtained by purposely positioning the nozzle of the paint spray gun very close to the samplers. The smallest filters (one-fourth of surface area) with the greatest velocities (7.31 cm/s) applied in this test allowed breakthrough of HDI. Results obtained from filter samples allowing lesser velocities and with the flow rates measuring 1 L/min, registered no HDI breakthrough. Furthermore, the results of this test proved that high airborne HDI concentrations are likely to occur in this type of CARC-spraying operation.

In the final stage of the field phase, 96 filter samples and nine impinger test samples were collected in a CARC paint spray booth and analyzed. The results from this test are presented in Table 5. The samples, denoted as A series, were analyzed according to the OSHA procedure, and all others (B and C series) by employing the newly developed technique for the removal of excess derivatizing reagent. The transport modes of Series B and C were also different. The comparison between the results from the A and C series showed equivalent performance of both the OSHA method and the new technique. In addition, there were no discernable differences when comparing B and C series results, thus indicating no HDI-urea losses occurred due to the shipping of samples in the filter cassettes.

The B and C series filters were impregnated each with a tenfold increased amount of 1-(2-P)P as opposed to the A series that contained only 0.1 mg of reagent per filter. No positive conclusions were drawn as to whether OSHA's original amount was sufficient to quantitatively collect all of the HDI present in the sampled air. Since the GC-impinger monitoring procedure was not part of the comparison study of the OSHA and OSHA-modified methods, the data from this perfunctory part of the test are not included in Table 5. As in the first field study, the GC-impinger results were consistently higher than those obtained from corresponding 1-(2-P)P filter samples. (See Field Test 1 and Table 3).

In the analyses of B and C samples using the new HPLC procedure, the HDI-urea peak eluted at approximately 13 min. Since the normally late-emerging oligomer and 1-(2-P)P peaks were disposed of by flushing to waste, samples could be injected in approximately

TABLE 5—*Results from third field study[a] (comparative sampling and analysis by OSHA's and modified methods).*

Test No.	A[b] OSHA Method, ppb	B[c] Modified Method, ppb	C[b] Modified Method, ppb
1	1.31	2.27	2.15
2	8.56	8.51	8.25
3	2.72	3.03	3.16
4	3.73	3.20	2.47
5	1.77	1.73	1.65
6	0.52	0.59	0.60
7	0.43	0.67	0.58
8	<0.15	0.17	0.17

[a] Each entry—mean of four samples.
[b] Filters shipped to laboratory in their cassettes.
[c] Filters transported to laboratory in desorbing solutions (in glass vials with crimped seals).

15-min intervals. In the case of the A sample series that were analyzed by the OSHA procedure, the presence of oligomer and 1-(2-P)P peaks necessitated long analysis time and restorative periods for the column to return to original conditions. This demonstrates, in practice, the advantage of the newly devised procedure for providing faster analysis.

Conclusion

The results of testing presented in this study prove the 1-(2-P)P filters to be a reliable, convenient method of collection for HDI. At the same time, the modification of OSHA's analytical procedure offers several advantages. In the first place, the mechanical removal of unreacted 1-(2-P)P from the HPLC system permits analysis without complications (deterioration of columns, etc.). This in turn allows for placement of larger amounts of 1-(2-P)P on filters to ensure capturing all HDI potentially encountered in high CARC-bearing atmospheres. Furthermore, the modified method greatly shortens the analysis time and, therefore, is more useful for routine environmental surveillance testing where large numbers of samples need to be handled.

Acknowledgment

The opinions or assertions contained herein are the private views of the authors and are not to be construed as reflecting the views of the U. S. Department of the Army or the Department of Defense.

References

[1] Rosenberg, C. and Tuomi, T., *Journal*, American Industrial Hygiene Association, Vol. 45, No. 2, 1984, pp. 117–121.
[2] Keller, J. and Sandridge, R. L., *Analytical Chemistry*, Vol. 51, No. 11, 1979, pp. 1868–1970.
[3] Sango, C., *Journal of Liquid Chromatography*, Vol. 2, 1979, pp. 763–774.
[4] Anderson, K., Gudehn, A., Levin, J. O., and Nilson, C. A., *Chemosphere*, Vol. 11, No. 1, 1982, pp. 3–10.
[5] Goldberg, P. A., Walker, R. F., Ellwood, P. A., and Hardy, H. L., *Journal of Chromatography*, Vol. 212, 1981, pp. 93–104.
[6] Langhorst, M. L., *Journal*, American Industrial Hygiene Association, Vol. 46, No. 5, 1985, pp. 236–243.
[7] Sango, C. and Zimerson, E., *Journal of Liquid Chromatography*, Vol. 3, 1980, pp. 971–990.
[8] Walker, R. F., Ellwood, P. A., Hardy, H. L., and Goldberg, P. A., *Journal of Chromatography*, Vol. 301, 1984, pp. 485–491.
[9] Chang, S. N. and Burg, W. R., *Journal of Chromatography*, Vol. 246, 1982, pp. 113–120.
[10] Burright, D., "Diisocyanates; Method 42," Organic Carcinogen and Pesticide Branch, Occupational Safety and Health Administration Analytical Laboratory, Salt Lake City, 1983, pp. 1–59.
[11] Marcali, K., *Analytical Chemistry*, Vol. 29, 1957, pp. 552–558.
[12] Esposito, G. G. and Dolzine, T. W., *Analytical Chemistry*, Vol. 54, 1982, pp. 1572–1575.
[13] Hastings-Vogt, C. R., Ko, C. Y., and Rayn, T. R., *Journal of Chromatography*, Vol. 134, 1977, p. 451.
[14] Bagon, D. A. and Purnell, C. L., *Journal of Chromatography*, Vol. 190, 1980, pp. 175–182.
[15] Hardy, H. L., *Journal*, American Industrial Hygiene Association, Vol. 45, 1984, pp. B30–B32.

Summary

Summary

Papers in this publication fall into four categories within the general areas of sampling and calibration: General Topics; Indoor Air Measurement; Ambient Air Measurement; and Workplace Atmosphere Measurement. The following is a summary of the papers for each category.

General Topics

This category includes two papers that discuss the basic principles of sampling and calibration that are important no matter what kind of measurements are made. *Kratochvil* emphasizes the fact that measurements are not made of an atmosphere but of samples considered to be representative of it. Such samples must be of unquestionable integrity and statistically related to the atmosphere of concern in order that valid scientific inferences can be made from the measurements. The paper goes on to present guidelines for the development of valid sampling plans and procedures to estimate the extent of sample variability.

The general principles of calibration are discussed by *Taylor* who reminds us that measurement is simply a comparison of unknowns with knowns that must match the former in all essential respects if valid data is to result. The paper discusses various approaches to calibration and ways to evaluate and minimize calibration uncertainty.

Indoor Air

An overview of indoor air monitoring is presented by *Levin*. Because this is an area of recent concern, there are many unsolved problems of sampling and calibration and each building monitored presents its own set of problems, due to specific design and occupancy considerations. The need for standardization to achieve comparability of data is emphasized.

The paper by *Fortmann et al.* discusses important details of monitoring of buildings, including the selection of sites, and the actual sampling and measurement procedures to be used. *Sterling et al.* discusses the important relation of ventilation to providing an acceptable environment and how air circulation can cause or aggravate contamination problems from both external and internal sources.

The important subject of sampling for microbiological contamination in indoor air is discussed by *Fradkin*. Consideration must be given to what and where to sample as well as how to sample. The latter is of special importance since improper sampling techniques can destroy or change the viability of organisms and vitiate the results.

Ambient Air

Although ambient air has been analyzed for many years, still there are unresolved problems of sampling and calibration resulting from the introduction of new measurement techniques, the need to measure exotic constituents, and the lowering of limits of concern of well-known constituents.

The first paper, by *Rook*, describes how estimates of precision and accuracy can be made from statistically analyzing calibration data. A paper by *Puzak and McElroy* discusses the important question of calibration traceability to the U.S. Environmental Protection Agency (EPA) for State and local laboratories. Unless such is achieved, the compatibility of measurement data obtained for national ambient surveillance and legal enforcement is open to questions of both technical soundness and legal defensibility.

Gases provided by specialty gas manufacturers provide the basis for calibration of most monitoring measurements. *Denyszyn and Sassaman* point out the precautions that must be observed to produce reliable mixtures and techniques that may be used to verify their limits of reliability.

Both theoretical and practical considerations in the use of permeation tubes to produce calibration gases are discussed by *Mitchell*. Various aspects of the preparation and verification of the composition of dilute gaseous mixtures are discussed by *Fried and Sams*. *Dorko and Hughes* describe various approaches that may be used to produce and verify the composition of reactive gas mixtures used to calibrate monitoring instruments.

Workplace Atmospheres

Workplace atmospheric sampling is conducted to investigate sources of contamination and to assess levels of exposure to industrial materials. The development of the strategy employed requires the cooperative input of industrial hygienists, physicians, toxicologists, analytical chemists, safety engineers, and process engineers. *Norwood* discusses all of this and presents a rationale for sampling based on traditional approaches with the use of statistical methods to aid in selection of sample size and data interpretation. The use of computer simulation in the design of sampling strategies is also discussed.

Melcher reviews the important question of laboratory and field validation of passive monitors. Information is presented for the selection of sorbents for pump and tube collection and for solvent and thermal desorbtion for analysis of collected material.

A comprehensive guide for industrial hygiene sampling, developed by the U. S. Army Industrial Hygiene Agency, presented by *Belkin and Bishop*, should be generally useful when measuring worker exposure. The important aspects of specificity and accuracy of detector tubes, used extensively for surveys of potential problems, are reviewed by *McKee and McConnaughey*.

Isocyanates have been sampled by impingers, solid sorbents, and coated-filter techniques. The paper by *Dharmarajan et al.* evaluates these techniques from the point of view of collection efficiency. The concluding paper by *Podolak et al.* discusses the collection and analysis of airborne hexamethylene diisocyanate, using impregnated filters. The importance of removing excess reagent from the coated filters, prior to analysis, is emphasized.

Final Remarks

Advances in analytical methodology have provided opportunities for better understanding of atmospheric phenomena and for the identification and control of contaminants. However, such measurements must be made reliably using advanced concepts of quality assurance and they are critically dependent on relevant samples and adequate calibration.

Sampling is a discipline in itself, involving knowledge of the subject area of concern, and measurement and statistical expertise. Because measurement is a comparison process, adequate calibration is essential, reliable calibrants must be used, and appropriate calibration procedures must be implemented, if measurement data are to be meaningful.

It is the hope of ASTM Committee D-22 and the authors of these papers that the 1985 Boulder Conference will have been useful in providing an increased awareness of the importance of sampling and calibration and in advancing the state of the art in these important areas of measurement.

John K. Taylor
Coordinator for Chemical Measurement Assurance and Voluntary Standardization, Center for Analytical Chemistry, National Bureau of Standards, Gaithersburg, MD; editor.

Author Index

Subject Index

Specificity, 176
Spray painting, 203
Stability of gaseous mixtures, 101
Standard Reference Materials
    (National Bureau of Stan-
    dards), 92
Standardization, 14
Statistical error, 146
Sulfur dioxide, 26, 87, 110
Survey questionnaire, 47

T

Thermal environment, 28, 46
Tight building syndrome, 46
Toxic materials, 166, 176
Trace gas calibration systems, 101,
    110, 121
Trace measurements, 101, 121
Tunable diode laser absorption spec-
    trometry, 121

U

Ultra-trace measurements, 101, 121
Uncertainty estimate, 81

V

Ventilation, 23, 28, 46
Viruses, 66
Volatile organic compounds, 26

W

Workplace atmospheres
  air conditioning, 203
  air movement, 28, 203
  building performance, 46
  building sickness, 22
  comfort factors, 46
  health factors, 46
  humidity, 28
  industrial hygiene monitoring,
    141, 149, 166
  overview, 21
  quality measurements, 35, 46
  sick buildings, 46
  solid sorbent samplers, 149
  thermal environment, 28, 46
  tight building syndrome, 46
  ventilation, 23, 28, 46